All-Embracing Manufacturing

International Series on
INTELLIGENT SYSTEMS, CONTROL AND AUTOMATION: SCIENCE AND ENGINEERING

VOLUME 59

Editor

Professor S. G. Tzafestas, National Technical University of Athens, Greece

Editorial Advisory Board
Professor P. Antsaklis, University of Notre Dame, Notre Dame, IN, USA
Professor P. Borne, Ecole Centrale de Lille, Lille, France
Professor D.G. Caldwell, University of Salford, Salford, UK
Professor C.S. Chen, University of Akron, Akron, Ohio, USA
Professor T. Fukuda, Nagoya University, Nagoya, Japan
Professor S. Monaco, University La Sapienza, Rome, Italy
Professor G. Schmidt, Technical University of Munich, Munich, Germany
Professor S.G. Tzafestas, National TechnicalUniversity of Athens, Athens, Greece
Professor F. Harashima, University of Tokyo, Tokyo, Japan
Professor N.K. Sinha, McMaster University, Hamilton, Ontario, Canada
Professor D. Tabak, George Mason University, Fairfax, Virginia, USA
Professor K. Valavanis, University of Denver, Denver, USA

For further volumes:
http://www.springer.com/series/6259

Gideon Halevi

All-Embracing Manufacturing

Roadmap System

Gideon Halevi
Industria and Management
Technion
Dubnov St. 20a
Tel Aviv, Israel

ISBN 978-94-007-4179-9   ISBN 978-94-007-4180-5 (eBook)
DOI 10.1007/978-94-007-4180-5
Springer Dordrecht Heidelberg New York London

Library of Congress Control Number: 2012934823

© Springer Science+Business Media Dordrecht 2012
This work is subject to copyright. All rights are reserved by the Publisher, whether the whole or part of the material is concerned, specifically the rights of translation, reprinting, reuse of illustrations, recitation, broadcasting, reproduction on microfilms or in any other physical way, and transmission or information storage and retrieval, electronic adaptation, computer software, or by similar or dissimilar methodology now known or hereafter developed. Exempted from this legal reservation are brief excerpts in connection with reviews or scholarly analysis or material supplied specifically for the purpose of being entered and executed on a computer system, for exclusive use by the purchaser of the work. Duplication of this publication or parts thereof is permitted only under the provisions of the Copyright Law of the Publisher's location, in its current version, and permission for use must always be obtained from Springer. Permissions for use may be obtained through RightsLink at the Copyright Clearance Center. Violations are liable to prosecution under the respective Copyright Law.
The use of general descriptive names, registered names, trademarks, service marks, etc. in this publication does not imply, even in the absence of a specific statement, that such names are exempt from the relevant protective laws and regulations and therefore free for general use.
While the advice and information in this book are believed to be true and accurate at the date of publication, neither the authors nor the editors nor the publisher can accept any legal responsibility for any errors or omissions that may be made. The publisher makes no warranty, express or implied, with respect to the material contained herein.

Printed on acid-free paper

Springer is part of Springer Science+Business Media (www.springer.com)

Preface

This book presents a new (improved) production management system.

The uniqueness of the system lies in its notions and tools, which are listed here in random order of importance:

- Routing is a variable
- The task of a process planner is to create a roadmap and not to create routing
- Production planning treats each order independently
- The system creates a *working* product structure based on level product structure
- Set critical order and give it priority in production planning
- The system eliminates or rectifies bottlenecks in production planning
- Shop floor control is maintained by resource searches for free operations.
- It enables alterations in production plans at any point in the process
- It serves as decision support system (DSS) to management

Theoretically, production planning and control is basically a very simple task.

A manufacturing plant receives orders for a product or a number of products, along with information such as quantities and delivery dates. The resources of the plants are known, the products are listed in a "Bill-of-Materials" (BOM) are known, and the routing is given. The task of production planning is to make sure that the orders will be ready on time in the most economical process. That's all.

Yet the commonly shared concept is that production planning and control is a very complex task. Over 130 complex production-planning methods have been devised over the years. Yet the search for "THE" method is ongoing.

This search has covered all sorts of ideas except one, *emulating modes of human thought*. Human thinking is very flexible; it allows setting of objectives and consideration of various methods to accomplish them, and when a disruption occurs it motivates alterations in original decisions in order to resolve the problem.

For example: If you want to go from point A to point B, you would probably study a map and plan the optimum route to take. This is a "present time" decision. However, at another time when you have to travel the same way, say at night, you might change the route; in winter you probably would look for a route with maximum shelter from precipitation. In summer you might choose a route that provides protection from the sun. In springtime you might choose a route with a particularly nice view. Despite the original routing decision, if you run into disruptions, such as a blocked road (bottleneck), multiple red traffic lights or heavy truck traffic, you might decide that, instead of waiting for the obstacle to clear, it is better to once again consult the road map and change the route in order to find a path with no obstacles. Such a change is possible at each junction. It might result in a longer route but it will be faster in time. The original decision would not prevent thinking human from adapting a new route.

Such a method may be used in production planning and scheduling. But why was it not even considered? Because there is no roadmap to consult for alternative routes, i.e., there is no GPS to provide guidance mid-process. The processing routing consists of fixed data and instructions. It was defined at some time in the past and is valid till some event calls for alterations.

The one who set the routing is the process planner, who is always very busy with generating routing for new items and cannot handle jobs that have already a routing but with minor order changes, such as quantity.

Furthermore, there are several criteria of optimizations that are affected by routing such as:

- Optimization of a single operation
- Optimization of an individual item
- Optimization of producing a product
- Optimization of producing a product mix
- Optimization of factory business

A single routing in a company's data base cannot accommodate all these criteria; it requires a roadmap method.

There was an attempt to develop a computer aided process planning (CAPP) program to generate routing from a drawing (CAD), but in spite of all the effort invested, this project did not materialized and the idea faded away.

Thus routing remains as a reflection of fixed data in the company's data base and cannot add flexibility to production planning. To overcome resource overload, in particular jobs competing for access to resources, the planner must decide which job will get priority. Other jobs must be shifted backward or forward in the schedule, which might mean increasing work in process or delay in order delivery.

1 This Book Presents a Method of Constructing an Important Tool: A Roadmap

1.1 Routing: Process Planning

Routing is the stage that transforms raw material into the form specified by the engineering drawing. This task should be carried out separately for each part, sub-assembly and assembly of the product.

Process planning is a decision making task for which the prime optimization criterion is to meet the specifications in the engineering drawings. The secondary criteria are cost and time with respect to the constraints set by company resources, tooling, know-how, quantity required, and machine load balancing. Some of these constraints are variable or semi-fixed; hence, the optimum solution obtained will be valid only with respect to those conditions considered at the time the decisions were made.

Therefore, under roadmap technology the process planner task is not to specify routing but rather to create a database (spread sheet) containing all possible process steps. Routing can be generated automatically in a split second, by the user taking into consideration the present (on-line) state of production.

The roadmap routing concept divides the task into three stages:

Stage 1: Technology stage; Generates BP—Basic Process. It is the "best" possible process from a technology standpoint. It does not violate any physical law. It is theoretical from a specific shop viewpoint.

Stage 2: Transformation stage; it construct an Operation—Resources matrix that lists all required operations, as generated by the BP. It searches the database to determine available resources and transform processing time of each operation in order to consider the restraints on each specific resource, and builds the content of the matrix ($T_{i,j}$—The time to process operation i on machine j).

Stage 3: Decision (mathematics) stage; Computes the path and sequence of operations that will result in the optimum routine according to the criteria of optimization. The matrix format represents almost an infinite number of possible processes.

Therefore the task of process planner is not to set routing but rather to generate a roadmap of possible processes, and let each department alter routing according to the current on-line manufacturing state.

A roadmap method of generating routing is independent of human process planners and each user may generate a routing suitable to the batch being processed, such as:

- Minimum materials cost
- Maximum production
- Maximum profit
- Minimum investment
- Minimum processing cost
- Medium quantity demand
- High quantity demand

This tool may resolve an overload, not by deferring job processing time but rather by selecting, when possible, different routing operations.

Roadmap routing generation programs have a feature that can respond to user requests, such as: select criteria of optimization; block resource; forced routing; alternatives; compare cost to processing time, etc.

1.2 Production Planning

The traditional approach to planning and execution regards routing as static and unalterable, therefore the planning is simple, but it robs the shop of production flexibility and efficiency.

The basic notions of hierarchical approach techniques are:

- Use the "best" routing for the job.
- Using the "best" routing for maximum production optimization will result in the shortest throughput.
- The larger the batch quantity, the higher the productivity.
- Releasing jobs based on MRP (or ERP) to the shop floor will assure maximum efficiency and adherence to delivery dates.

These notions appeared to be logical and were accepted without justification. However, subsequent research led to the conclusion that these notions are in many cases incorrect.

The term 'best" routing is obscure; the roadmap may generate over 20 routings for each job, each one to serve a different objective.

Using maximum production criterion of optimization routing for scheduling a random product mix showed that this criterion resulted in the longest throughput time compared to other criteria. It was not academic research but it shed reasonable doubts on this notion.

Some research conclusions on scheduling rules indicate that:

- Mean flow time is minimized by using SPT sequencing dispatching rules.
- Mean inventory is minimized by using SPT sequencing dispatching rules.
- Mean waiting time (prior to start processing) is minimized by using SPT sequencing dispatching rules.
- Maximum lateness is minimized by using SPT sequencing dispatching rules.

2 SPT Means: Shortest Processing Time

If SPT is so overwhelming in a particular setting, the routing can be designed to be composed of several short processing time elements instead of a few long operations.

Another method of keeping SPT is keeping batch quantity low. The quantity is defined in each specific order, but at least one should not increase the batch quantity by combining jobs of other orders.

To summarize the above discussion, some of the most important all-embracing manufacturing technology notions for production planning are as follows:

- Regard routing as a variable.
- Treat each order, with its product structure, individually.

- Convert a level-product structure into a working-product structure on a time scale.
- The product with the earliest start time of a lower level item is regarded as the critical order.
- Give priority to this critical order for stock allocation (after each allocation the critical order might be changed.
- Load for processing each order as a unit rather than combining items.
- Release early scheduled periods to shop floor for further processing.

Carrying out the planning actions as described above, results in:

- Minimum processing lead time
- Meeting delivery dates
- Maximum resource utilization
- Minimum work in process
- Minimum capital tie down in production
- Elimination of bottlenecks in production

2.1 Shop Floor Scheduling and Control

The objective of shop floor planning and control is to ensure that the released jobs for a period will be completed on time and in the most economical way possible. To achieve this goal, total flexibility is required.

The process spread sheet establishes a network of possible routings while deferring the decision of which path to take to a later stage. Furthermore, the decision of which path to take can be changed after each technological operation.

The proposed shop floor control approach is based on the concept that whenever a resource is free, it searches for a free operation to perform. A *free resource* is defined as a resource that has just finished an operation and the part has been removed, or is idle and can be loaded at any instant. A *free operation* is defined as an operation that can be loaded for processing at any instant. An example would be the first operation of an item for which the raw material and all the auxiliary jobs are available, and is within reach of the resource loading mechanism.

2.2 Management Decision Support System

All-embracing manufacturing technology is basically an engineering driven system that creates a processing tool for the use of management. Its main feature is the ability to generate routing and scheduling online without the need for the presence of a process planner during the processing; therefore, it may assist management by supplying objective information and simulations needed to make decisions from engineering data, such as:

- Resource planning
- A company's level of competitiveness

- Improvement of level of competitiveness
- Expansion of manufacturing capabilities
- Introduction of new manufacturing technologies
- Assisting in establishing delivery dates
- Assisting in setting optimum selling prices

Any simulation the manager requires can be accessed by the appropriate computer program, without the need to consult with a process planner.

The book contains six chapters, which presents two stages of the manufacturing process.

The engineering stages of how to generate variable routing are presented in Chap. 1. The production planning stage is presented in Chap. 2. A system demonstration is in Chap. 3.

The management stages are regarded as innovative tasks and depend on designer creativity. Therefore they are not considered as part of the technology. Each manager may direct his or her company by his or her own methods and preference.

However, to demonstrate the benefits that one may acquire by appreciating the potential of using the proposed technique, Chaps. 4, 5 and 6 are included.

Contents

1 Introduction 1
   1   Introduction 1
           1.1   Industrial Management 4
   2   All-Embracing Manufacturing Technology 6
           2.1   Human Emulation Examples 7
           2.2   System Notions 7
           2.3   All-Embracing Technology Concepts 8
   3   All-Embracing Technology System Architecture 9
           3.1   Basic System Files 9
           3.2   Master Management: Management Information Generator 12
           3.3   Production 13
   4   Summary 15

2 Process Planning: Routing 17
   1   Introduction 17
   2   First Stage: Process Planning 19
           2.1   Stage 1 Example 23
   3   Second Stage: Tranformation 24
           3.1   Preliminary Resource Selection 25
           3.2   Operation Transformation 26
           3.3   Computational Method 30
   4   Stage 3: Routing Generator 30
           4.1   Definition of the Combinatorial Problem 33
           4.2   General Matrix Solution 36
   5   Conclusion 44

3 Production Module 45
   1   Introduction 45
           1.1   Traditional Approach 46
   2   Production Management Strategy: Roadmap Manufacturing 51
           2.1   Roadmap Notions 51
           2.2   Production Planning 52

		2.3	Stock Allocation Priority	53
		2.4	Stock Allocation Method	55
		2.5	Adjust Quantities	58
		2.6	Capacity Planning: Resource Loading	59
		2.7	Job Release for Execution	62
	3	Shop Floor Control		63
		3.1	Concept and Terminology	64
		3.2	Algorithm and Terminology	66
		3.3	Summary	70
4	**Production Planning: Demonstration**			**73**
	1	Introduction		73
		1.1	The Scenario	74
	2	The Planning Steps		79
		2.1	Determination of Stock Allocation Priorities	80
		2.2	Stock Allocation	81
		2.3	Capacity Planning: Resource Loading	85
		2.4	Job Release for Execution	94
		2.5	Shop Floor Control	95
5	**Product Specifications and Design**			**101**
	1	Introduction		101
		1.1	Product Design: Engineering Design	101
		1.2	Design Goals: Task Specifications	103
		1.3	Product Specifications Methods	108
		1.4	Master Product Design System: Concept Design	119
		1.5	Master Design System: Detail Design	123
		1.6	Summary	131
6	**Detail Design**			**133**
	1	Introduction		133
		1.1	Assembly Steps	138
	2	Assembly-Oriented Planning		139
		2.1	Manual Assembly	141
		2.2	Automatic Assembly	141
		2.3	Robotic Assembly	141
		2.4	Hybrid Automatic-Manual Assembly	143
	3	Design Constraints for Assembly		144
		3.1	Design Rules	144
		3.2	Orientation	146
		3.3	Fastening	147
	4	Component Design for Placement		149
		4.1	Component Which is Nearly Identical on Both Sides	149
		4.2	Headed Fasteners	149
		4.3	Components Design for Placement	149
	5	Summary		150

7	**Management Decision Support System**		151
	1 Introduction		151
	2 Plant Performance Measurement		153
		2.1 Reasource Suitability to Products	154
		2.2 Production Planning Performance Level	155
		2.3 Shop Floor Performance Level	156
	3 Resource Planning		156
		3.1 Resource Recommendation Module	158
	4 Maximum Profit Criterion of Process Planning Optimization		166
	5 Determining a Process for Maximum Profit		168
		5.1 First Stage	169
		5.2 Second Stage	170
		5.3 Third Stage	170
		5.4 Testing the Algorithm	171
		5.5 Summary of Maximum Profit	172
	6 Determining Delivery Date and Cost		172
		6.1 Generating Alternatives of Cost: Delivery Date: New Order	173
		6.2 Summary	175
		6.3 Chapter Summary	176
Index			179

Chapter 1
Introduction

Abstract "All-embracing Manufacturing Technology" is a system that introduces flexibility and agility to production planning. Properly implemented, it automatically dissolves constraints, bottlenecks, and disruptions while maintaining simplicity of procedures. In this setting, no decision is finalized ahead of its time for execution in the plan. Among other features, the system considers the present state of a company's total inventory of orders and up-to–the-minute conditions on the shop floor. It considers technology to be the dominant tool of manufacturing planning. It is a computer-driven system that strives to emulate the behavior of a thinking human being.

1 Introduction

Theoretically, production planning and control is a very simple task. The plant receives orders that define the product or products required, their quantities and desired delivery dates. The resources of the plants are known; the product bill of material and the routing are known. The task of production scheduling is to make sure that the orders will be ready on time for shipping, that's all.

It seems strange that in order to fulfill this simple task, over 130 complex production planning methods have been proposed over the years. Yet the search for "THE" method is ongoing.

A common shared excuse is that production planning is a complex task. The complexity is due to the vagaries of a dynamic environment such as: a machine fails; a tool breaks; employees are missing; orders change; parts are rejected and reworked; power failures occur; there might be a mismatch between load and available capacity, unrealistic promised delivery dates etc.

Analysis of the reasons for this complexity shows that:

- Engineering decisions are being made at a too early stage. Some decisions are made several years before their execution time. Therefore, they cannot conform to the dynamic situation of the manufacturing process.
- Most of the disruptions (that cause the complexity) are a result of stiffness of systems in which such early decisions are made.
- Planning and execution regards routing as static and irreversible, therefore the planning robs the shop of production flexibility and efficiency.
- Data transfer between disciplines consists of decisions already made and does not reflect the decision maker's intentions, alternative methods, or new ideas.
- Engineering stages are mostly human oriented activities and rely on planners' experience and intuition. Therefore, it is possible that sophisticated mathematical production management models are working with biased and questionable data.
- The immediate linkages between jobs and orders are fragile.
- The hierarchical approach. The design is based on a top-down approach and strictly defines the system modules and their functionality. Communication between modules is strictly defined and limited in such a way that modules communicate only with their parent and child modules. In a hierarchical architecture, modules cannot take initiative; therefore, the system is sensitive to perturbations, and its autonomy and reactivity to disturbances are weak.
- The criteria of optimization are not always synchronized with the total objective.
    - Local optimization of a single operation does not necessarily lead to optimization of the item.
    - Item optimization does not necessarily lead to optimization of the product.
    - Product optimization does not necessarily lead to optimization of the product mix.
    - Product mix does not necessarily lead to optimization of the business.

The basic notions of hierarchical approach techniques are:

- Use the "best" routing for the job.
- Using the "best" routing for maximum production optimization will result with the shortest throughput.
- Productivity increases the larger the batch quantity. Therefore it considers all orders equally, injects the orders into its product tree, and combines a quantity of individual items, when possible.

These notions are in many cases wrong and are certainly inadequate to deal with complexity on a sound foundation.

The major conventional objectives of production planning are:

- Meeting delivery dates.
- Keeping capital tied down in production to a minimum.
- Minimizing manufacturing lead time.
- Minimizing idle time on resources.

In this traditional setting, jobs are aggregated and scheduled over currently existing (or routinely planned) resources, thus are subject to unexpected overloads, bottlenecks, etc.

As routine plans are static and irreversible in real time, problems have to be solved by setting job priorities. The first job gets access to the existing resource and the other jobs are moved forward or backward looking for resources that might be available at different points of the timeline.

Such solutions contradict the objectives of a smooth flow, and this method can result in an increased WIP load, lengthening of throughput time, and risking failure to meet delivery dates.

Since the traditional methods do not supply solutions, new methods and solutions are needed. One philosophy behind the proposed solutions is:

PRODUCTION IS VERY COMPLEX. Therefore we need more, and more complex, computer programs and systems to regulate and control it.

Such proposed systems recognize the fact that *constraints* exist and offer methods to overcome them. The method might be by mathematical tools such as: genetic algorithms; multi-grade fuzzy approaches; shift constraints; Mimetic algorithms; etc.

Other developing methods are: Theory of Constraints (TOC); just-in-time (JIT); reactive scheduling; Drum-Buffer-Rope (DBR); optimized production technology (OPT); hybrid flow shop scheduling etc.

Another philosophy is:

PRODUCTION IS VERY COMPLEX. Therefore there is no chance of success in attempting to build a system that will solve all problems simultaneously. Hence the role of computers should be limited to supplying information and leaving decisions to human intelligence.

Several methods are described in this proposal: total quality management (TQM); agile manufacturing; agent-driven approach; enterprise resource planning (ERP); holonic manufacturing system; etc.

This trend envisions the computer as a secretary, a tool that can store and retrieve data and present it to the user on request. Thus it allows a human to make decisions.

Another philosophy is:

PRODUCTION IS VERY COMPLEX. Therefore the only way to make production systems more effective is to simplify them.

The simplification can came by considering constraints as obstacles that have to be resolved by technological means rather than by mathematics or other abstract procedures. Most constraints are due to the stiffness and notions of the relevant system. A constraint can be eliminated before it becomes an obstacle.

We must remember, of course, that computers can be a working tool; they can automatically execute decisions made previously by humans. In the manufacturing process, each decision is a direct result of a previous stage, such as: An order file and product structure table can check the status of inventory, search in the purchasing file to retrieve supplier files, select supplier and issue an order, check quality control and the validity of the received goods, enter the items into the production planning files, etc. None of these steps needs a human involvement.

1.1 Industrial Management

Manufacturing constitutes only one branch in the organization chart of an industrial enterprise, albeit a dominant one, since it controls the daily activities of the other disciplines. However, it represents only one aspect of the activities of industrial management. Management must consider all the activities in the enterprise. The objectives of management are:

- Implementation of the policy adopted by the owners or the board of directors.
- Optimization of return on investment.
- Efficient utilization of assets such as workers, machines and money.

In other words, industry must make profits.

Therefore, the optimization criteria for management decisions must be cost, capital tie-down in production, and profit; these are collectively referred to as finance criteria.

However, by present day methodology, not even a single stage of the manufacturing process considers finance, economics and cost as their primary objective. Each stage optimizes its task to the best of its ability. Each stage in the manufacturing cycle according to its function has its own objectives and criteria of optimization. Even if each stage functions optimally, this does not necessarily guarantee overall optimum success with respect to management's prime objectives.

The traditional manufacturing cycle is a one-way chain of activities, where each link has a specific task to perform, the previous link being regarded as a constraint.

Thus, for example, master production schedules accept the routing and bill-of-materials as fixed data (as well as quantities and delivery dates); they do not question these data and their planning must comply with them.

A process planner accepts product design and its bill-of-materials without question; in fact, the planner does not even consider a product as a whole, but rather regards the production of each part as a specific task. Only if problems are encountered in defining the process for a particular part does the planner turn to the product designer and suggest or ask for a change in design.

The capacity planner accepts the routing as fixed data, and employs sophisticated algorithms to arrive at an optimum capacity plan.

Therefore, the chain of activities that comprises the manufacturing cycle is considered as a series of independent elements having individual probabilities of achieving a criterion. The probability of success of any link is independent of every other link with which it is functionality associated.

Thus the overall probability of the chain optimally achieving a particular criterion is:

$$Pj = P_{j1} \times P_{j2} \times P_{j3} \times \ldots \times P_{jn} = \prod_{i=1}^{i=n} Pj,i$$

1 Introduction

Table 1.1 The manufacturing stages (links) and their criteria

Link i=	Stage	Criterion j
1	Engineering Design	$P_{1,1}$ = Performance
		$P_{2,1}$ = DFM, DFA ...
		$P_{3,1}$ = Finance
		.... $P_{n,1}$
2	Process Planning	$P_{1,2}$ = Performance
		$P_{2,2}$ = Time
		$P_{3,2}$ = Finance
		.... $P_{n,2}$
3	Methods, Time and Motion study	$P_{1,3}$ = Time
		$P_{2,3}$ = Ease of
		$P_{3,3}$ = Finance
		.... $P_{n,3}$
4	Master production schedule	$P_{1,4}$ = Load profile
		$P_{2,4}$ = Delivery dates
		$P_{3,4}$ = Finance
		.... $P_{n,4}$
5	Material Requirement - Resource Planning	$P_{1,5}$ = Delivery dates
		$P_{2,5}$ = Load profile
		$P_{3,5}$ = Finance
		.... $P_{n,5}$
6	Capacity Planning	$P_{1,6}$ = Due dates
		$P_{2,6}$ = WIP (Work in Process)
		$P_{3,6}$ = Finance
		.... $P_{n,6}$
7	Purchasing	$P_{1,7}$ = Due dates
		$P_{2,7}$ = Quantity and Quality
		$P_{3,7}$ = Finance
		.... $P_{n,7}$

where Pj = the overall chain probability of achieving criterion j,
Pji = the probability of achieving criterion j in link i.

The seven stages of the manufacturing cycle that are subject to the finance criteria are shown in Table 1.1; the other stages need not be considered.

As can be seen, finance criteria appear in all stages as the third criterion, but not as the primary criterion. If we assume an 80% probability of achieving the finance criterion in each of these seven stages, then the overall probability of the finance criteria being optimally achieved in the manufacturing cycle will be

$$0.8^7 \times 100\% = 0.21\%.$$

It should be noted that the ability to predict each probability Pj,i is difficult at best; therefore, discussion is qualitative only.

Management must have tools (controls) in order to achieve its objectives. One of these tools is the budget. However, the budget is based on engineering data, the bill-of-materials and routing. Thus it can only assure that the outcome will not be worse than the planning; it cannot improve the planning.

Another tool is the operation organization of the company. A company is organized with respect to key functions, each function being concerned with a different aspect of the operation and fighting for its point of view.

For example: Sales would want to be able to promise early delivery and competitive prices and thus would favor high level of inventory and low cost production.

Finance would prefer a minimum amount of capital tie-down in production and thus would favor a low level of inventory and short lead time production.

A production manager would emphasize that all work stations must have jobs and thus would favor a high level of in-process inventory and long lead times.

Only if each of the disciplines stands up for its own interests will a good balance in the overall operation of the plant be reached.

Value engineering is another tool that management can use to examine and improve the various manufacturing activities. It will usually be used when providing support to a particular product in respond to market demands. Although value engineering is an important tool, it is seldom employed as a part of the manufacturing cycle; if it were to become a standard part of the manufacturing cycle, it would be part of the "Establishment" and probably cease to serve its original purpose.

Standardization and simplification are additional tools that can be used to improve the financial aspects of manufacturing. However, they are administrative measures without real control.

Another approach that management may take is to focus on profit opportunities rather than on efficiency.

Management must have tools in order to exercise control over operations. Manual tools are not perfect, and, consequently, all-embracing manufacturing technology can assist management in performing its task in an approved manner.

2 All-Embracing Manufacturing Technology

All-embracing manufacturing technology claims that:

PRODUCTION IS VERY SIMPLE and flexible by nature.

All-embracing manufacturing technology introduces flexibility and agility to production planning, it dissolves constraints, bottlenecks, and disruptions automatically while restoring simplicity. No decision is made ahead of time, but only at the time of execution. Therefore, it considers the present state of company's orders and of the shop floor. It introduces technology as the dominant part of manufacturing. It is a computer oriented system, imitating what practically any of us does in our personal lives.

2.1 Human Emulation Examples

For example: If you wish to go from point A to point B, than you would be well advised to study a map and plan the optimum route to take. This is a present time decision. However, when at another time you may have to travel the same way, say, at night, you might change the route; in the winter you probably will look for a route with maximum shelter from precipitation. In summer you might choose a route that protects you from the sun. In the spring time you might choose a route with a nice view.

Despite the original routing decision, if you run into disruptions such as a blocked road (bottleneck), red traffic light or a crowded road, you might decide that instead of waiting it is better to consult the (GPS) road map and change the route in order to find a path with no obstacles. Such a change is made at each junction. It might be a longer route but it will be faster. The original decision must not prevent one from adapting the new route.

Another example: when you drive a car and you see a red light (stop sign) at a distance ahead, you will start slowing the car. Or if you see a green light at a short distance you may speed up in order to pass through it.

There is not much sense in rushing in order to wait; or to drive normally in order to wait. The same reasoning should apply to production scheduling. One may reduce cutting speed in case the resource will have to be idle before the following job will be ready. Or one may increase cutting speed (if it is technically allowed) in order to save idle time for the next job.

Another example: It was decided to make for dinner dessert a cheesecake. A cheesecake is a dessert consisting of a topping made of soft, fresh cheese on a base made from biscuit, pastry or sponge cake. The topping is frequently sweetened with sugar and flavored or topped with fruit, nuts, fruit flavored drizzle and/or chocolate. If you want the topping to be nuts, and you do not have nuts, you may decide not to make the cake or to choose an alternative with whatever you have, such as chocolate.

All three parts are being made simultaneously and must be synchronized to be ready when needed. Suppose that for some reason the base is not ready there is no need to follow the original time scheduling and you might give priority to other waiting home tasks.

2.2 System Notions

All-embracing technology treats the manufacturing process as one interactive problem starting from engineering design to product shipment. It considers the manufacturing process as a nucleus and satellites rather than as a chain of activities. The engineering activities are the nucleus, and the other activities are the satellites. Thereby all-embracing technology introduces engineering, value engineering, and technology into all phases of the manufacturing process. It broadens the scope of alternate solutions and eliminates the artificial constraints used as an interface between the engineering and production phases in the chain of activities. Thus a better optimum solution can be reached.

Production planning today regards the engineering phases of manufacturing, that is, product design (bill-of-materials), process planning, and methods (routing), as unaltered. These individual phases create artificial constraints that exist only because the product designer, the process planner, and the production engineer do not (and in practice cannot) communicate continuously and work as a team.

The engineering problems in manufacturing are being transformed into mathematical problems at too early a stage, so that in spite of impressive mathematical techniques and a true optimum calculation, from a mathematical standpoint, a real optimum solution is not always obtained.

2.3 All-Embracing Technology Concepts

The objectives of all-embracing technology are to ensure customer satisfaction while increasing productivity and reducing manufacturing costs. These objectives are the same as those of many other systems. The difference lies in the approach and concepts employed. The all-embracing technology approach makes use of the following notions:

- There are infinite ways of producing a work piece.
- The cost and lead time required to produce a component are functions of the process used.
- There are infinite ways of meeting design objectives.
- In any component about 75% of the dimensions (geometric shape) are nonfunctional (fillers). These dimensions can vary considerably without affecting the component performance.
- The cost and lead time required to produce a component are functions of its design. A minute change in fillets or dimensions to suit a standard tooling or an existing set up on a machine can result in significant cost variations.
- With present-day techniques, competition between jobs over resources will always occur. The method and logic of resolving this competition, that is, pull forward or backward, defeat the main purpose of production planning, which is to meet delivery due dates and minimize the amount of capital tied down in production.
- There exists a manufacturing optimum that is theoretical from a specific shop standpoint, but practical from a technology standpoint.
- Optimum is a contingent expression; one must specify an Optimum of: a single operation; an item; producing a product (several items); producing a product mix; factory business.
- Production planning decisions should be made at the moment of need, according to the actual state on the shop floor by a self-organization system

The basic philosophy of all-embracing technology is that all parameters in the manufacturing process are flexible, that is, any of them is subject to change if such change contributes to increased productivity. In such a flexible and dynamic environment, the only constant parameters are the products to be manufactured and the resources available at the shop. Product objectives are external to the manufacturing cycle and must be preserved. Change recommendations are welcome, but should not be carried out automatically; they must be approved by management.

All-embracing technology can suggest and evaluate different manufacturing environments. Manufacturing costs, capabilities, and deviation from the theoretical optimum can be computed for any environment and submitted as information to management. Management, according to its forecasts and financial considerations, can reach an intelligent decision as to the desirable manufacturing environment.

3 All-Embracing Technology System Architecture

All-embracing technology manufacturing cycle architecture is divided into three main modules, **Basic system files; Master Management; Production Management**, as shown in Fig. 1.1.

Each one of the modules transfers information to other modules while within its own sphere, all disciplines are connected to one another by exchanging data, ideas, and information.

3.1 Basic System Files

This module includes product design, process planning and a product structure.

Product design and process planning are the two most important tasks of the manufacturing process. They account for over 80% of the processing cost and 30% of the lead time. These two are interrelated and affect one another; therefore they should work together in a concurrent method. A product has to seduce the customer, with its options and appearance.

3.1.1 The Purpose of Design

The purpose of the design is to transform the objective set by management into a detailed set of engineering ideas, concepts and specifications. Engineering design theories are employed to translate the objectives into engineering specifications. Many ideas and concepts will be formulated and analyzed, and the best conceptual solution will be determined. This conceptual design will define separate engineering tasks of lower level (detail design) until the last detail of the design is decided upon.

The optimization criteria for the decisions made in the design stage are for the most part engineering considerations such as: weight, size, stability, durability,

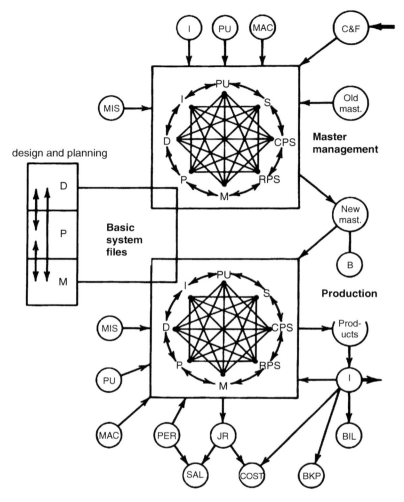

Fig. 1.1 All-embracing technology manufacturing cycle. Notation, *D* product design, *P* process planning, *M* methods, time, and motion study, *RPS* requirement planning, *CPS* capacity planning, *S* shop, *I* inventory, *PU* purchasing, *MAC* machine file, *C&F* customer orders and forecasting, *MIS* miscellaneous, *B* budget, *PER* personnel, *JR* job recording, *SAL* salaries, *COST* costing, *BKP* bookkeeping, *BIL* billing

ease of operation, ease of maintenance, noise level, cost, and so on. Some of the criteria conflict with each other, and thus the decision will often be a compromise. However, the designer's primary criterion in making a decision is to meet the product objectives. This is the designer's most important responsibility, since errors in production are not as critical as errors in design. To be on the safe side, the designer will tend to incorporate as many safety factors as possible.

3 All-Embracing Technology System Architecture

The design decisions reached in the engineering design stage are transformed into a set of detailed engineering drawings and product structure (part lists). It is an editing process, constrained by the explicit rules and grammar of engineering language, namely drawings.

Product specification as well as product design are innovative tasks and depend on designer creativity. This book presents methods to increase creativity, and arrives at a product design that will result in low cost; ease of manufacturing and ease of assembly. Design techniques such as: group technology (GT); design for manufacturing (DFM), design for assembly (DFA), the gate method, axiomatic design, decision support system (DSS), may assist the designer in this task.

3.1.2 The Purpose of Process Planning

The purpose of process planning is to transform raw material into the form specified and defined by the engineering drawing. This task should be carried out for the assembly, and separately for each sub-assembly and individual part of the product. This stage is basically analogous to the engineering design stage, but here the nature of the objective is different.

Process planning is a decision making task for which the prime optimization criterion is to meet the specifications given in the engineering drawings. The secondary criteria are cost and time with respect to the constraints set by company resources, tooling, know-how, quantity required, and machine load balancing. Some of these constraints are variable or semi-fixed; hence, the optimum solution obtained will be valid only with respect to those conditions considered at the time of making the decisions.

Process planning and design are completely independent tasks, but in many cases an insignificant change in the design may reduce significantly the process plan cost and lead time. Therefore, there must be communication between these important tasks.

Process planning is a decision making task for which the prime optimization criterion is to meet the specifications in the engineering drawings. The secondary criteria are processing time and cost with respect to the constraints set by company resources, tooling, know-how, quantity required, and machine load balancing. Some of these constraints are variable or semi-fixed; hence, the optimum solution obtained will be valid only with respect to those conditions considered at the time the decisions were made.

Therefore, under all-embracing technology the process planner task is not to specify routing but to create a roadmap (database) containing all possible process steps and let each user generate automatically a routing that is appropriate to its application.

Process planning as well as product design is an innovative task and depends on designer creativity. This book will present methods to increase creativity, and to arrive at a process that will result in low cost, ease of manufacturing and ease of assembly. Design techniques such as: GT; CAPP; RCAPP may assist the planner in this task.

3.2 Master Management: Management Information Generator

Master production planning is a coordinating function between manufacturing, marketing, finance, and management. The objective of the master production schedule is to supply management with a tool for controlling and a "look ahead," tool which is necessary in order to plan the future of a company. It provides simulation of capacity requirements for different marketing forecasts, purchasing of new equipment, and profit or loss forecast. It indicates the necessary requirement planning with respect to shop floor space, warehouse space, transport facilities, and manpower.

The objectives specify what products are to be produced, in what quantity, in what dates and at what cost. The sources of these data can be confirmed customer orders, forecasts of future demands, or a combination of both. These are non-engineering sources and are not concerned with plant capacity. In the master production schedule, the impact of alternate production plans on plant capacity and load balance is assessed. The result is a practical master production plan, which is the basis for manufacturing activities, as well as management and finance activities.

The budget is the conversion of manufacturing activities to cost. The future development of the company—for example, its predicted profit, investments in addition of new resources, and in research and development, is reflected in the budget and the master production plan.

Industrial management is responsible for the success of the enterprise; it controls all disciplines' activities. The objective of management is, as mentioned above:

- Implementation of the policy adopted by the owners or the board of directors.
- Optimum return on investment.
- Efficient utilization of workers, machines and money.

All-embracing technology does not pretend to advise management on how to meet these objectives or to run their business.

The objective of this module is merely to serve as a Decision Support System (DSS). All-embracing technology will generate, on request, reliable data, and will supply information for management in order to simulate different strategies and make decisions. Some all-embracing technology topics are:

- Resource planning.
- Human requirement planning.
- Cash flow profit forecasting.
- Product selling price.
- Economic lot size.
- Maximum profit.
- Plant layout.
- Order logbook.

3.3 Production

Production's task is to plan and produce products according to management orders and policy. The most important conventional objectives of the task are, as previously listed:

- Meeting delivery dates.
- Keeping capital tied down in production to a minimum.
- Minimizing manufacturing lead time.
- Minimizing idle time of resources.

The traditional approach to planning and execution regards routing as static and unalterable; therefore it runs into scheduling problems that lead to items competing over resources.

One method to solve this competition is to set priority rules and let one of the competitors be processed and move the others forward or backward to periods where the resource is idle. This method may increase work in process and result in delay in meeting delivery dates.

An additional method is to increase processing quantities by combining items from different orders to be processed collectively. Doing so increases processing time and might cause delay in delivery dates, and it makes production planning a complex task as the relationship of an item to its order is disrupted.

All-embracing technology with its roadmap system proposes to:

- Treat each order, with its product structure, individually. It does not attempt to increase quantity by combining similar items into one processing batch.
- Avoid having items compete for resources by choosing substitute routing and thus eliminate such competition.
- Set critical order and let this order be the first to be treated. Therefore disruptions are solved automatically by the common sense of the system.

All-embracing technology performs this task in the following steps:

- Determination of stock allocation priorities.
- Stock allocation.
- Adjustment of quantities by economic considerations.
- Capacity planning—machine loading.
- Job release for execution.
- Shop floor control.

3.3.1 Determining Stock Allocation Priorities

Determination of stock allocation priorities is done by transforming the level-based product structure of all open orders into a time-based product structure.

The starting date is the order delivery date which is reduced by the time required to process each item as calculated by: the quantity required multiplied by the process time as indicated in the spread sheet prepared by the process planning module. The order with a lower level item as its early start date is stated as the critical order.

3.3.2 Stock Allocation

The available stock is assigned to the critical order. After each allocation the time-based product structure is updated to the new state and the process repeats till all orders are considered. A new critical order might be declared.

At the end of the stock allocation step, the working product structure includes only those items that have to be produced, or purchased. The working product structure is not similar to the master product structure, as some items might be missing altogether, others might have a different quantity.

3.3.3 Capacity Planning: Resource Loading

The objective of capacity planning is to determine what job to release to the shop floor for execution. Capacity planning plans all the jobs to be carried out in the plant in order to meet all customers' required delivery dates. It is long range planning. The actual processing is done on the shop floor which requires a short term production plan.

Capacity planning strategy is to operate in down-top mode i.e. start with the first operation of the critical low level item of the critical order. This item has priority. Loaded operations are recorded on a spread sheet format which indicates loading status. If the preferred resource is occupied, a look-ahead feature will calculate the time till it will be free. In case that the waiting time is too long, then the system uses its scan function looking for an alternate resource.

3.3.4 Shop Floor Planning and Control

The objective of shop floor planning and control is to make sure that the released jobs for a period will be completed on time and in the most economical way possible. To achieve this goal, total flexibility is assumed.

The proposed shop floor control approach is based on the concept that whenever a resource is free, it searches for a free operation to perform.

A *free resource* is defined as a resource that has just finished an operation and the part was removed, or is idle and can be loaded at any instant.

A *free operation* is defined as an operation that can be loaded for processing at any instant. An example would be the first operation of an item for which the raw material and all the auxiliary jobs are available, and is within reach of the resource loading mechanism.

4 Summary

This book's objective is to introduce the concept of flexibility and to demonstrate that it is a realistic and profitable approach.

Manufacturing is by nature a very simple task and not complex. Many of today's problems that make manufacturing complex are solved by elimination, that is, they are eliminated by introducing a new degree of freedom into the manufacturing cycle.

The main tool is the concept of a process planner task, which is to generate a roadmap of process planning options, and not to decide which routing is the best for the job. The routing should be a variable.

The concept of *variable routing* has an effect on manufacturing planning and control systems. Good ideas used to be rejected on the grounds that they are impractical. With today's advanced technology, they are now practical.

The second concept is that production planning should handle each order as a unit without combining items of several orders to be processed as one batch.

Chapter 2
Process Planning: Routing

Abstract The purpose of process planning is to set a routing that will transform raw material into the form specified and defined by the engineering drawing. It is generally based on human skill and intuition. A routing, once set, may be revised only by a process planner. This robs the manufacturing system of its natural flexibility.

The point of view of this book is that routing is a mathematical problem in combinatorics that does not require the expertise of a process planner. Therefore the task of the process planner is not to set routing but rather to generate a roadmap of possible processes, and let each discipline generate routing according to the immediate on-line manufacturing state.

Such an approach makes manufacturing simple and efficient.

1 Introduction

Process planning is an engineering task that determines the manufacturing cost and the efficiency of plant operations. This holds true for all types of manufacturing. Engineering tasks are generally based on human skill and intuition. In mass-production or chemical processes a lengthy study is made in order to formulate an optimum process. Such studies are not practical in batch-type manufacturing, since the cost of the study is often greater than the possible savings in manufacturing. It is, therefore, doubtful whether optimum processes are currently being employed in batch-type manufacturing.

This situation is particularly critical in the metal-cutting field, where about 75% of all items are produced in batches of 50 or less and where over 80 billion dollars are spent each year in the United States.

The purpose of process planning is to set a routing that will transform raw material into the form specified and defined by the engineering drawing. This task should be carried out for the total assembly, and separately for each sub-assembly and individual part of the product.

Process planning is a decision making task for which the prime optimization criterion is to meet the specifications in the engineering drawings. The secondary criteria are cost and time with respect to the constraints set by company resources, tooling, know-how, quantity required, and resource load balancing. Some of these constraints are variable or semi-fixed; hence, the optimum solution obtained will be valid only with respect to those conditions considered at the time the decisions were made.

The process planning and design are completely independent tasks, but in many cases an insignificant change in the design may reduce significantly the process cost and lead time. Therefore, there must be communication between these important tasks.

Process planning produces information for management and production. Management may request information in order to make decisions regarding resource planning, order due dates, etc. Such requests may come at random points in time and need up-to-date data. On the other hand, production cannot function without process planning. Routing is the basic conduit of information to production planning and scheduling. Preparing routing for each job order considering quantity, resource available, etc. consumes time, which might not be economical, therefore in many cases the same routing might be employed for several jobs regarding specific job details.

Process planning is human-oriented activity, highly dependent on individual skill, human memory, reference manuals and above all experience. To evaluate the efficiency of routing proposed by a human process planner a study was conducted. Drawings of parts having different complexities were given to several process planners of different backgrounds, and they were asked to define the process. It was amazing to note that the number of different processes defined per part coincided with the number of process planners participating. By the proposed routing, one may successfully determine the type of plants the process planner used to work and gain expertise. I am almost sure that anyone, anywhere, who conducts such a study, will get similar results. The process planner has to consider in any decisions certain components: parts, tools, resources, fixtures, technology, batch quantity and cost. The planner makes a good intuitive decision, based on past experience, but not aware of the controlling parameters. Therefore when one of the parameters changes, it is not always taken into consideration in generating a process plan.

Process planning is a decision-making process. The process planner has to specify which resources to use, which operations to perform on each resource, their sequence, and which tooling and cutting conditions to use. The sequence of decisions taken by the process planner introduces constraints. Once the process planner makes a decision, it becomes a constraint to the following decisions.

For example: A selected resource imposes constraints on the power available for a cutting operation; the torque at the spindle, the maximum depth of cut, the maximum cutting speed and the available speeds and feeds, the machining dimensions, the number of tools that can be used, the accuracy, the handling times. A single machining operation can be adjusted to comply with these constraints. But, machining cost and time will be a function of the selected resource. Similarly, a selected tool imposes constraints on the maximum cutting speed, depth of cut, feed rate and tool life.

It is accepted that these constraints are artificial ones; they are in effect only because of the sequence of decisions taken. Selecting another resource will impose different constraints. A different sequence of decisions will result in a different process plan.

These constraints are artificial, since they are dependent on the sequence of decisions. Thus the efficiency of a routing proposed by a manual process planner is questionable.

The roadmap (matrix) concept approach was developed to overcome dependence on the sequence of decisions, and the artificial constraints problem. To overcome this problem and comply with dynamic requirements, the proposed algorithm uses a roadmap concept and divides the process planning task into three stages:

Stage 1: Technology stage; Generates BP—Basic Process. It is the "best" possible process from a technology stand point. It does not violate any physical law. It is theoretical from a specific shop view point.

Stage 2: Transformation stage; constructing an Operation—Machines matrix. It lists all required operations, as generated by the BP. It considers the available resources, and transform processing time of each operation to consider each specific resource's constraint, and builds the content of the matrix (Ti,j—The time to process operation i, on machine j).

Stage 3: Decision (mathematics) stage; matrix solution. Compute the path and sequence of operations that will result in the optimum process plan according to the criteria of optimization. The matrix format represents almost an infinite number of possible processes. For a matrix of $N = 10$ operations and $M = 10$ resources, the number of process combinations is $N!M^N = 3.6288*10^{16}$. To find the appropriate process becomes a mathematical problem and not a technological one.

By the roadmap concept a process is generated using only real constraints (technological ones) by assuming an imaginary resource and tools. Thus a basic process (BP) is generated. The difference between conventional process planning and the roadmap process planning is that the process planner generates a process using an imaginary resource i.e. a resource that poses only technological constraints. This process can be generated by computer program or by manual effort.

2 First Stage: Process Planning

The first stage is engineering and is limited only by engineering technology. It disregards the available plant resources but it is practical from the engineering standpoint, thus all real strength and technical constraints have been taken into consideration. One may regard it as an imaginary resource with unlimited capabilities of power, speed, feed, accuracy (or the ability to imagine the best resource on the market, anywhere in the world). The process planner has to build a table with the following format as shown in Table 2.1. Notice that the table does not specify a resource data but rather required specifications.

The data for *General part data* is retrieved directly from the part drawing. The columns are:

Operation type: List of process types such as:

- Milling
- Drilling
- Sanding
- Cleaning
- Riveting
- Welding
- Bending
- Forging

Priority—A priority value indicates the operation number that must precede the present operation. Priority 0 means that an operation can be the first resource operation or can be performed anywhere. Priority value other than zero indicates the operation number that must precede the present operation.

Relation—This operation must be done with the indicated operation without removing the part from the resource or fixture.

Due to technological constraints, such as geometric tolerances, some operations must be processed without removing the part from the resource or fixture. For example, a concentricity, or parallelism cannot be accomplished economically if the part is subject to two separate fixtures or resources. The value in this column indicates such a relationship between operations.

Operation data: The process planner examines the part drawing and makes his decision as to which processing type to select. This decision is affected by the size and complexity of the part, and by process capabilities of the plant. The method is universal process planning.

The operation details data are functions of the process type selected. For example the data for plastic injection moulding should include the following:

- Locking force
- Maximum pressure
- Maximum injection speed
- Injection volume
- Maximum temperature
- Die size

The data for metal cutting should include the following:

- Material type and code
- Material size
- Type of operation (external, internal, surface, etc.)
- Type of cut (rough, finish, chamfer, etc.)
- Length of cut

Table 2.1 Universal process planner decisions table

General part data				Computational data						
Seq no.	Operation	Priority	Relation	Tool Dia	Length mm	Depth mm	Feed mm/min	Speed m/min	Power KW	Time min
010										
020										
030										
040										
050										

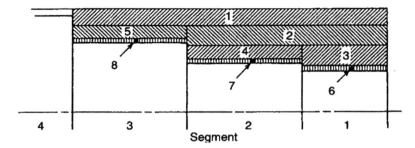

Fig. 2.1 Operation sequence as specified in BP stage

- Depth of cut
- Tolerance
- Surface finish
- Geometric tolerances
- Maximum feed rate
- Maximum cutting speed
- Depth of cut

Computational data depends on the general part data decision. Its purpose is to compute the data that will be needed to generate alternatives. Which operation type user might select the appropriate parameters?

For example: the length of cut and tool size are derivatives of the operation type and can be retrieved from the part drawing. Depth of cut value is dictated by the drawing. It is the difference between raw material and the top surface of the part (An = Am-Ad). However, the surface finish dictates the maximum depth of cut possible Ap. If An > Ap than an intermittent cuts pass should be added. Therefore decision as to feed rate must be assigned first, and then the depth of cut dealt with. A semi-finish cut might be required to bridge the depth between the finish and the rough cut. Cutting speed is a function of the part drawing and tool selection and can be retrieved from an appropriate table of cutting tools manufacturing. The operation power can be computed by using the previous parameters of the table.

The operations in the matrix are arranged in a random order. However, some operations must precede others. Figure 2.1 shows eight operations as they appear in the matrix. It is clear that operation 2 cannot be performed unless operation 1 is done and clears the way for the tool. Similarly, operations 3 and 6 must be made in that order. However, operation 2 clears the way for the tool that processes operation 5, therefore operation 5 may be processed after operation 2. Operation 8 may be processed after operation 5 and operation 7 after operation 4.

The Priority column indicates the constraints on the actual sequence of machining operations. Priority 0 (zero) means that this operation may be the first machining operation, or be processed anywhere. Priority value other than zero indicates the operation number that must precede the present operation.

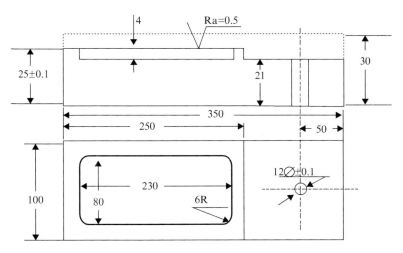

Fig. 2.2 Sample part

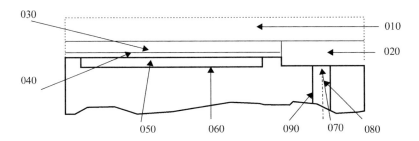

Fig. 2.3 Detail process operations

2.1 Stage 1 Example

For demonstration, a sample part is shown in Fig. 2.2. The process planner considers only technological constraints and ignores order quantity and available resources.

After examining the drawing, the process planner decided to choose a milling operation type of surface and pocket and drill for the hole. Noticing that the top surface is marked as surface finish of $Ra = 0.5$, this surface must have a minimum depth of cut; however, since 5 mm of material has to be removed, the process should be done with three cutting passes; 0.2 mm finish cut; 0.4 mm semi-finish cut; and 4,4 mm rough cut. (Note: there is a limit to depth of cut size before finish cut). The operations are marked in Fig. 2.3 and are recorded in the table.

The first operation, rough milling with face mill diameter 125 mm (100 mm part size + tool overlap, and select closest tool available). The length of cut is 378 mm (350 mm part size + 28 mm tool entrance).

The second operation, rough milling, can be done only after operations 010 due to tool entering. (This operation could be done as the first operation, priority 0 but

Table 2.2 Process operation table

No.	Operation	Priority	Tool Dia	Length mm	Depth mm	Feed mm/min	Speed m/min	Power KW	Time min
010	Rough milling	0	125	378	4.4	808	100	20	0.47
020	Rough milling	010	125	128	4.6	735	100	20	0.17
030	Semi-finish milling	020	125	278	0.4	905	148	2.2	0.31
040	Finish milling	030	125	378	0.2	200	165	0.39	1.89
050	Rough pocket milling	010	80	150	4.0	1,093	102	20.6	0.24
060	Finish pocket milling	050	12	472	0.4	120	24	0.33	4.16
070	Center drill	020	3	3	–	0.05	14	0.025	0.03
080	Twist drill	070	7	21	–	0.16	15.7	0.3	0.22
090	Core drill	080	12	21	–	0.19	23.5	0.5	0.20

The feed rate and the speed of cut values are taken from tables or equations, and are recorded in the table; these two values with material tables can compute the operation power required

in such case the depth of cut should have to be 9 mm instead of 4.6 mm). Selecting a tool of 125 mm, the length of cut (cross side) is 128 mm (100 mm + 28 mm). Depth of cut is $(30 - 21 - 4.4) = 4.6$ mm.

The semi-finish milling, operation 030 can be performed after the rough operation, thus priority 020 due to tool entrance. The finish cut 040 can be performed after the semi-finish cut, thus priority 030.

The pocket may be resourced right after the rough cut 010 thus its priority is 010, and its finish cut after it, thus operation 60 priority is 50.

Processing the hole requires three operations in sequence; Center drill twist drill and core drill and is started right after its surface is resourced (operation 020).

Table 2.2 demonstrates a process operation table for the sample part. Naturally the basic process time as presented in Table 2.2 is a theoretical one as it does not consider the resources of a specific plant. The operation time is not mandatory in this table. It is computed by dividing the length of cut by the feed rate. It is given for reference in the next stage.

The number in the priority column indicates that the relevant operation can be resourced only after the indicated operation number in the priority column has been carried out. It does not have to follow this operation immediately. This means that the sequence of operations may be: 010; 020; 030; 040; 050; 060; 070; 080; 090; **OR** 010; 020; 030; 070; 080; 090; 040; 050; 060; **OR** 010; 050; 020; 070; 030; 080; 040; 090; 060; **OR** any other combination.

3 Second Stage: Tranformation

The first stage, considers only engineering and technological constraints. Thus, a theoretical process is generated, theoretical, that is, from a specific shop point of view (e.g. such a resource may not be available), although practical from an engineering point of view, since all technical constraints have been taken into account.

For each individual operation, the parameters have been defined: depth of cut, feed rate and cutting speed etc. These values guarantee that the part will be produced economically and meet all specifications as defined in the drawing. However, it does not consider the available resources in a given industrial shop, nor the other operations.

The problem at the second stage is to transform the basic operations into practical ones from a specific shop point of view, i.e. to adjust the process operations to the available resources. This is basically a mathematical problem where all the alternatives available in terms of resources have to be generated and computed. The only technical knowledge required is how to make the specific operation comply with a particular resource specification. These adjustments can easily be constructed. Thus the problems of resources selection and process planning are transformed from an engineering problem into a mathematical one.

The second stage is to construct a two-dimensional matrix: Operations—resources and computes the processing time for each junction of the matrix. It is basically a mathematical problem, where all data can be generated and computed. The only technical knowledge required is how to make the specified operations comply with a particular resource constraint.

This stage handles technological data. However, as the equations for transformation are straightforward, a computer program can easily be developed to perform this task.

To consider individual resource capabilities and constraints, the basic operation time is translated and adjusted to comply with each individual resource specification. It is obvious that the processing time cannot be decreased; it may only be increased. The adjustment considers the following factors: resource physical size, resource accuracy, special features, available power & torque, available speeds and feeds, number of tools, type of controls, handling time etc.

3.1 Preliminary Resource Selection

In order to lessen the number of alternatives and thus reduce solution time, a preselecting of resources will be made. This preselecting should be made so as not to affect the optimum solution.

3.1.1 First Stage in Resource Selection

The available resources are handled one by one. Initially the physical size of the resource is checked. In the case that the resource cannot accommodate the part it is excluded from further consideration.

Next, the resource accuracy and type are checked. In the case that a resource cannot perform even a single operation, that resource is excluded from the matrix.

Otherwise the resource remains in the matrix. The time or costs of the specific operations that cannot be performed is set to high values (99). This will prevent the selection of this resource for that operation, while leaving the possibility of selecting this resource for other operations. Old, inaccurate, or low cost resources may be used for rough operations, while they are not suitable for finishing operations.

3.1.2 Second Stage in Resource Selection

A resource whose power is lower than the minimum required does not stand a chance of being selected for any operations. Therefore, it can be excluded from further consideration without affecting the optimum process plan, unless there are no other resources.

A resource with more power than required has no advantage over lower power resources. Therefore, such a resource can be excluded from further consideration, unless its hourly rate is lower than the low-power resource or it has a spindle speed higher than that of the lower-power resource and this spindle speed is required by one of the operations.

3.1.3 Third Step in Resource Selection

This third step is employed only if the number of remaining resources is too high. An estimation of a resource's chance of being chosen serves as the basis of this step. The chances are estimated by comparing the resource time or cost for the given resource with those of other resources.

- Select the three resources that have the lowest total machining time (single resource solution for the job).
- Select resources that have a minimum value in any single operation.
- If there are still too many resources, select those that have a minimum value in the higher number of operations.

Assume that six resources are being considered. A short list of specifications of these resources is given in Table 2.3.

3.2 Operation Transformation

Following are three simplified explanations of how operation transformation may take place. The user may choose any one of them according to the desired quality and simplicity.

3 Second Stage: Tranformation

Table 2.3 Specifications of available resources

Resource Number	Resource specifications	Power KW	Speed RPM	Handling Time min	Relative cost $
1	Milling Machining Center	35	1,500	0.10	4
2	Large CNC milling	35	1,200	0.15	3
3	Manual milling resource	15	1,500	0.66	1.4
4	Small drill press	1	1,200	0.66	1
5	Old milling resource	15	2,400	1.0	1
6	Small CNC milling	10	3,000	0.25	2

3.2.1 Relational Method

As long as resource accuracy, type, and size comply with the operation requirements the operation can be adjusted to specific resource constraints and specifications. Initially, the maximum values of depth of cut, cutting speed, and feed rate are compared to those of the proposed process and when needed are reduced to the available value as necessary.

For minor changes in depth of cut, an attempt is made to decrease the rough cut and increase the depth of cut, if the surface finish allows this maneuver. The maximum depth of cut (a) for the finish cut may be computed by equation

$$a_{max} = (32Ra)/BHN^{0.8}$$

where Ra is the surface finish in micrometers and BHN is the material hardness (Brinell number).

The difference between a_{max} and the depth of cut for the finish cut is the amount of the allowable increase and decrease of the finish and rough cut respectively.

If such adjustments of the depth of cut are not sufficient, the rough cuts are split into a rough cut and finish cut or a rough, semi-rough and finish cut. An attempt is made to use the maximum depth of cut allowed for the resource for the rough cut and minimum for the finish cut.

If there are several operations on the same segment, as may be indicated by the operation field, their sum is considered as the required depth of cut, and a split of cut is made by applying the same logic as used for the case of one operation with an oversize depth of cut.

If the required feed rate exceeds the available rate, it is reduced to the maximum available feed rate.

The controlling values of a resource are torque and power. Both are functions of the cutting forces. The basic process considered the allowable forces due to part bending, tension and compression, chucking counteractive forces, jigs and fixture forces etc. In the adjustment of the operation cutting conditions the cutting forces may be decreased but never increased. Therefore, there is no need to check their value. However, the cutting forces are a major factor in computing the torque and power requirements, which must be adjusted to those of a specific resource.

Computation of the adjusted cutting force (F) depends on how the basic cutting force was established.

If an expert indicated the cutting force the following equation can be used to adjust the value:

$$F2 = F1 * (a2/a1) * (f2/f1)^{0.75}$$

where $F2$ = the adjusted cutting force, $F1$ = the basic cutting force, $a2$ = the adjusted depth of cut, $a1$ = the basic depth of cut, $f2$ = the adjusted feed rate, and $f1$ = the basic feed rate.

If an expert did not specify the cutting force, it may be computed for turning operations by the following equation or any other equation chosen by the user.

$$F = Cp * a * f^{0.75} * (0.16 * HBN^{0.35})$$

where Cp = a constant that can be taken from a machinability material databank, or the value of 225 may be used for steel, 140 for iron, 365 for high temper alloys, and 130 for nonferrous alloys; a = adjusted depth of cut, f = the adjusted feed rate; and HBN = material hardness (Brinell number). Equations for milling, grinding, etc. may be easily found in textbooks.

Torque is a function of cutting forces multiplied by the diameter, while power is a linear function of cutting forces and cutting speed. Therefore, the adjustment of the torque will be handled first. In the case that the computed torque exceeds the resource torque, the feed rate will be reduced down to 57% of its recommended value. If this reduction is not sufficient, a reduction in depth of cut will take place (in the case of milling, an attempt to change tool diameter and/or number of teeth will be made before splitting the depth of cut).

As the cutting speed is a function of depth of cut and feed rate, the basic cutting speed is recomputed before checking the available cutting speeds (RPM). If the required RPM exceeds the available one, it is reduced to the maximum available RPM.

Based on the modified depth of cut, feed rate, and cutting speed, the required power is computed. If the available power is lower than the required power, the cutting speed is reduced to 63% of its initial value.

If this reduction is not sufficient, a module is used to determine the most economic measure to reduce the required power (i.e. split depth of cut, reduce feed rate or reduce cutting speed or a combination of all).

The adjusted cutting conditions are used to compute the machining time (Ti,j) of the specific operation(s) on the specific resource. Handling time is added to the computed machining time and the sum will be entered into the matrix.

This feature of constructing the matrix enables one to evaluate the economics of using conventional resources versus CNC resources, manual operations versus tool-assisted operations, old resources rather than new resources etc.

To convert time to cost, the specific resource hourly rate will be multiplied by the machining time.

3.2.2 Set of Rules Method

When machining power is less than that required, the cutting speed or the feed rate or the depth of cut has to be reduced. (P = power; T = time; V = cutting speed; f = basic feed rate; 1 = basic data; 2 = reduced data.)

Cutting Speed Reduction

Cutting speed reduction will proportionally reduce power and increase cutting time.

$$P1/P2 = T1/T2,$$
$$T2 = T1 * P2/P1,$$
$$V2 = V1 * T2/T1.$$

Feed Rate Reduction

Feed rate reduction approach is to maintain the cutting speed and reduce the feed rate to a value $f2$.

In such a case, the machining power and time is:

$$P1/P2 = T2/T1 = (f1/f2)^{1-y}.$$

Since by definition, $f1$ is always greater than f2, T_2 will always be greater than T_1.

This means that it is more profitable to reduce the cutting speed than to reduce the feed rate. Reducing cutting speed will also result in an increase of tool life.

A practical conclusion is that power adjustment should be made according to the following priorities:

If the power has to be reduced by less than 50%:

> Reduce the cutting speed down to its lower value; or
> Reduce the feed rate down to its lower limit value; and
> Reduce the depth of cut, i.e. split the cut into more than one pass and adjust cutting speed and feed rate.

If the power has to be reduced by more than 50%:

> Split the depth of cut to more than one pass and then use the priorities above.

A more sophisticated measure may include a change in tool cutting edge angles and in the number of teeth in a milling tool also. The rules in such cases are:

- Down by 85%—reduce cutting edge angle;
- Down by 70%—as above, plus a reduction in cutting speed;
- Down by 60%—as above plus a reduction in feed rate;

- Down by 51%—split depth of cut to two passes;
- Down by 40%—split and reduce cutting speed; split depth of cut.

Note: a reduction of cutting forces is not effected by reducing the cutting speed. Use only the reduction of feed rate as a first attempt and if this is not enough, split the depth of cut.

Remember that if the cutting forces and the power have to be reduced, handle first the reduction of cutting forces, and check to ensure that the power problem still exists.

3.3 Computational Method

Let the process planner enter the data using his or her own skill or use the attached flowchart. See Fig. 2.4. Operation transformation module (2 pages).

The converted time values are shown in Table 2.4.

For a low-level quantity the table shows that the best routing is to use resource 1 for all operations. Note that the difference between 1 and 2 is insignificant and using resource 2 may be alternative routing.

The time matrix can easily be converted to a cost matrix. This is done by multiplying the time by the hourly rate. Thus the value of $T_{i,j}$ is converted to $C_{i,j}$, where $C_{i,j}$ represents the cost of performing operation i on resource j.

The converted time to cost values are shown in Table 2.5.

The minimum cost column value gives the processing cost of each individual operation, and its resource number. Total minimum cost column value gives the optimum cost, as each individual operation is processed by the best resource. However, individual operation optimization does not result in part optimization. To arrive at part optimization, expenses (penalty) for setup, inspection, transfer time between resources, etc. must be considered. Note that these expenses are for a batch and therefore depend on the quantity to be produced. The part optimum process will be given by a path that is a compromise of all the operations and results in a minimum sum of processing cost value registered in the matrix, with sequence considered by the priority number and with the addition of a penalty for changing resource.

The criterion of optimization can be maximum production or minimum cost, depending on the content of the matrix.

4 Stage 3: Routing Generator

The process planner task is to construct an Operation–resources matrix as discussed in the previous section. It is not the planner's task to generate routings. Each user according to needs and time of need may construct a routing automatically by using an appropriate algorithm.

4 Stage 3: Routing Generator

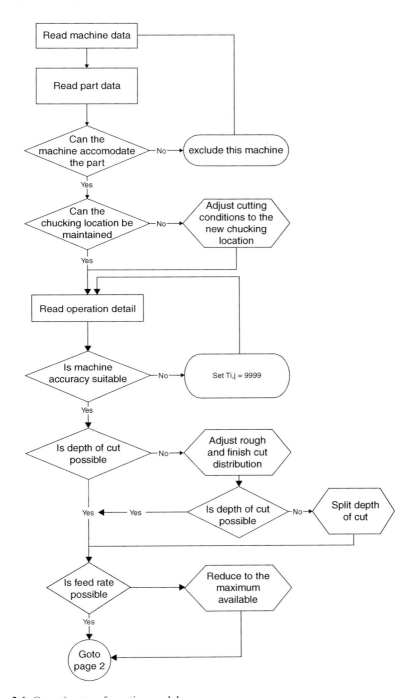

Fig. 2.4 Operation transformation module

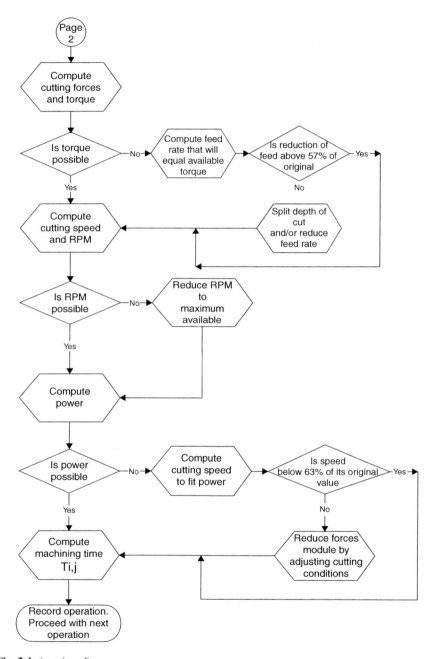

Fig. 2.4 (continued)

4 Stage 3: Routing Generator

Table 2.4 Resource—operation time matrix

Operation	Time min	Priority	Rel	R #1	E #2	S #3	OU #4	R #5	CE #6
010	0.47	0	0	0.57	0.62	1.28	99	1.62	1.19
020	0.17	010	0	0.27	0.32	0.88	99	1.22	0.59
030	0.31	020	0	0.41	0.46	0.97	99	99	0.56
040	1.89	030	0	1.99	204	2.55	99	99	2.14
050	0.24	010	0	0.34	0.39	0.99	99	1.32	0.74
060	4.16	050	0	4.26	4.31	4.38	99	99	4.41
070	0.03	020	0	0.13	0.18	0.69	0.69	1.03	0.28
080	0.22	070	0	0.32	0.37	0.88	0.88	1.22	0.47
090	0.20	080	0	0.30	0.35	0.86	0.86	99	0.45
Total	7.69			8.59	9.04	13.92			10.82

Table 2.5 Resource—operation cost matrix

Operation	Time min	Priority	Rel	R #1	E #2	S #3	OU #4	C #5	CE #6	Minimum cost
010	0.47	0	0	2.28	1.86	1.79	99	1.62	2.36	1.62 M5
020	0.17	010	0	1.08	0.96	1.23	99	1.22	1.22	0.96 M2
030	0.31	020	0	1.64	1.38	1.36	99	99	1.12	1.12 M6
040	1.89	030	0	7.96	6.12	3.57	99	99	4.28	3.57 M3
050	0.24	010	0	1.36	1.17	1.39	99	1.32	1.48	1.17 M2
060	4.16	050	0	17.04	12.93	6.75	99	99	8.82	6.75 M3
070	0.03	020	0	0.52	0.54	0.97	0.69	1.03	0.56	0.52 M1
080	0.22	070	0	1.28	1.1	1.24	0.88	1.22	0.94	0.88 M4
090	0.20	080	0	1.20	1.05	1.20	0.86	99	0.90	0.86 M4
Total	7.69			34.36	27.12	19.50		21.64	17.45	

The problem is to search for the best sequence of operations and the best sequence of resource utilization. This is basically a combinatorial problem which leads to an unlimited number of alternatives that have to be evaluated in terms of economic efficiency. The following sections will suggest solutions to this problem.

4.1 Definition of the Combinatorial Problem

At this stage optimization of processing a part is considered and not the individual operation, i.e. taking into consideration time for part transfer from resource to resource.

To illustrate this problem, let us examine the resource power required for processing a part as shown in Table 2.2, it is: 20; 2.2; 0.39; 20.6; 0.33; 0.5 KW. It seems that there is quite a wide spread of required resource power. In case that a resource with the maximum power is selected (20.6 kW) then operations that require less power may be processed on that resource but its processing time will remain unchanged. Note: Op. 010 is a rough operation that requires power, but not accuracy. It takes power of 20 KW and only 0.47 min. While Op. 040 is a fine cutting operation, it requires an accurate resource and takes power of 0.39 KW and 1.89 min. It does not make sense to use the same resource for both operations.

In case that a resource with lower power is selected (for example 12 kW) operations that require higher power than is available may be processed, but the processing time will be increased.

There is a wide selection of resources that can produce the part with respect to its power, however there is no compromise regarding accuracy. An inaccurate resource cannot produce an accurate part, even with added processing time.

If the 'best' resource for each individual operation is selected, additional time has to be added covering transfer time (and cost) between resources. However, if the sequence of operations can be altered, savings in transfer time may be realized.

It is clear that transfer time and cost are functions of the required quantity. For low quantities, the best compromise might be to process the part with only one resource. This might increase the direct machining time, but reduce the transfer time. For very high quantities, however, the best compromise might be to process each operation with its best resource, which will increase transfer time and reduce direct processing time. For moderate quantities, the best compromise is not so clear. Calculations should be made to find the point when direct processing time and transfer time should be increased, to arrive at the minimum total processing time (or cost). To assist in finding the solution regarding quantity, a matrix of all combinations is to be created.

Operation optimization must not result in Part optimization. To arrive at part optimization, expenses for setup, inspection, transfer time between resources, etc. must be considered. Notice that these expenses are for a batch and therefore depend on the quantity to be produced. The part optimum process must be a compromise between all expenses (i.e., transfer penalty and the individual operation cost). The part optimum process will be given by a path that is a compromise between the operations and results in a minimum sum of processing cost values registered in the matrix, with a sequence constrained by the priority number and with the addition of a penalty for changing resources.

There are two criteria for optimization: maximum production and minimum cost. The process generated for each criterion is completely different from the other one. The process recommended for the maximum production criterion results in the shortest processing time but ignores processing cost, and the process recommended for the minimum cost criterion results in the lowest processing cost but ignores processing time.

Comparing the two optimization criteria for the part given in Tables 2.4 and 2.5 indicates that for maximum production, the processing time is 8.59 min and the cost for example 34.36, while for minimum cost the cost is (the sum of minimum cost in Table 2.5) 17.45 and the time is (the sum of the time in Table 2.4 of the minimum cost resource indication) 11.27 min.

For an example of constructing a maximum production routing form, the time matrix (Table 2.4), which is supposed to meet maximum production criteria of

4 Stage 3: Routing Generator

Table 2.6 Routing with sequence change

Operation	Time min	Priority	Rel	M M1	A M2	C M3	H M4	IN M5	E M6	Minimum cost
010	0.47	0	0			1.79		1.62		1.62 M5
020	0.17	010	0		0.96					0.96 M2
050	0.24	010	0		1.17					1.17 M2
030	0.31	020	0						1.12	1.12 M6
040	1.89	030	0			3.57				3.57 M3
060	4.16	050	0			6.75				6.75 M3
070	0.03	020	0	0.52					0.56	0.52 M1
080	0.22	070	0				0.88		0.94	0.88 M4
090	0.20	080	0				0.86		0.90	0.86 M4
Total	7.69				2.13	12.11			3.52	**17.76**

optimization, it is clear that the best resource may handle all operations in short processing time, and the routing seems to be simple.

However, while using it in scheduling it might result in being the longest throughput, as probably all operations will select this resource and thus create a long queue. Calling for minimum cost routing might result in a shorter throughput time and lower cost as the operation will use resources that are underloaded.

The following demonstrates how routing for minimum cost may be constructed.

Suppose that a quantity of 1,000 pieces is ordered and the setup cost and other expenses to resource the batch are $200. Thus a penalty for transferring the job from one resource to another is $200/1,000 = \$0.20$.

Notice that, contrary to the maximum production criterion, the best resource for each individual operation varies. This means that six penalties have to be added to the processing cost.

Thus the minimum cost routing calls for six penalties and the routing cost is

$$17.45 + 6 * 0.2 = 18.65.$$

However, if the sequence of operations can be altered, the number of resources may be decreased and thus the number of penalties. Moreover, in case that the difference of processing cost/time is lower than the penalty, a further decrease can be made of the number of resources that are required to produce a part. Table 2.6 demonstrates that the number of resources may be reduced from 6 to 3.

The processing cost increased from 17.45 to 17.76 but it calls for only three resource changes.

Thus the total cost is $17.76 + 3*0.2 = 18.36$.

Naturally the routing is a function of the quantity and the penalty. In case of a small quantity the penalty is high and the routing will be to use one resource, the one with the minimum total time. Refer to Table 2.5 to use only resource 3.

For a very large quantity the penalty will be almost zero and the selected routing is as shown in Table 2.6.

The process planning method, as described, is systematic and probably does not differ much from the process planner's way of thinking (probably unconsciously).

It ensures better and more consistent routing. It results in part optimization rather than operation optimization. It solves the problem of artificial constraints (sequence of decisions).

4.2 General Matrix Solution

Routing is a technological decision-making process. By the matrix concept it has been transformed from a technological decision into a mathematics decision. The definition of the mathematical problem is as follows:

Given: Operation (i)—Resource (j) matrix listing all operations and the process value for each operation on each resource ($V_{i,j}$). A decision is required as to which resource to use, which operation(s) to perform on each resource, what their sequence should be. The constraints are indicated by Priority—which indicate operations that must precede others and certain relationships that must tie operations together to be performed on the same resource.

Extra expenses and time should be added to cover extra setup, chucking, and transfer of parts between resources, additional complication in capacity planning and job recording and inspection etc. These extra expenses are called "penalty". Thus the penalty for a batch is a function of the quantity to be produced.

The larger the quantity the lower the penalty, and thus a higher profitability of selecting the best resource for each specific operation. Naturally, in each case the sequence of operations might be different.

The optimization criterion is either maximum production, or minimum cost. Figure 2.1 shows an eight operations sequence as they appear in the list.

Thus the operation sequence of 1, 2, 3, 6; 1, 2, 4, 7; and 1, 5, 8 must be maintained, even though any number of operations can be inserted in-between. Another consideration is tool access. This requirement restricts the sequence of the operations shown to 1, 2, 3, 4, 5, 6, 7, 8 or 1, 2, 3, 5, 4,7, 6, 8 or 1, 2, 5, 3, 4, 8, 6, 7 while operations 6, 7, 8 can come later.

The number of combinations of a matrix containing N operations and M resources is:

$$N! * M^N$$

This is an almost infinite number of combinations and it is impossible to use normal methods to solve this problem. (For $N = 10$ AND $M = 10$ the number of combinations is $3,828,800 * 10^{10} = 3.8288 * 10^{16}$). Therefore a special method has been developed to solve this problem in a finite time. The general solution is based mathematically on Bellman's theory of dynamic programming technique.

> "The basic feature of dynamic programming is that the optimum is reached stepwise, proceeding from one stage to the next. An optimum solution set is determined, given any conditions in the first stage. This optimum solution set from the first stage is then integrated with the second stage to obtain a new optimum solution, given any conditions. Then, in a sense ignoring the first and second stages as such, this new optimum solution is integrated

Fig. 2.5 Dynamic programming procedure

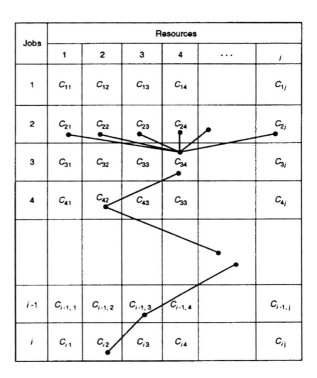

into the third stage to obtain still further optimum solutions and so on until the last stage. It is the optimum solution that is carried forward rather than the previous stage." This dynamic programming procedure is shown in Fig. 2.5.

The following example demonstrates this logic. At an intermittent point of the series of decisions it was decided that job 3 is to be performed on resource 4 (see Fig. 2.5). At this stage we can ignore how and why we reached this decision. The problem at hand is: to where should we proceed from this point, that is, on which resource should we perform job 2. This problem can be expressed mathematically as finding which of the following expressions has the minimum value:

$$T_{2,1} = T_{3,4} + C_{2,1}$$
$$T_{2,2} = T_{3,4} + C_{22}$$
$$T_{2,3} = T_{3,4} + C_{2,3}$$
$$T_{2,2} = T_{3,4} + C_{22}$$

. . .

. . .

. . .

$$T_{2,j} = T_{3,4} + C_{2,j}.$$

This is a finite problem and can be solved easily and fast. The number of combinations to be solved by this method is

$$N \times M.$$

Bellman's theory is based on two assumptions:

1. The total value is an accumulation of the individual job values. This assumption is true in our case, with exception of transfer time.
2. The total value at any point is independent of the path by which it was arrived at. This assumption does not hold in our case. Since the jobs (operations) are interdependent and their sequence may or must be changed in order to reach optimum, the path is one of the decisions to be reached by the solution.

It seems clear that in order to solve the problem at hand, a resource-operation matrix must be used. A discrete solution must be composed.

In the problem at hand the stages are referred to as jobs (operations) and decisions are made by choosing the optimum path between any two jobs. However, since the sequence of operations listed in the matrix is not fixed, this sequence can be changed. One of the problems to be solved is which sequence of operations will result in an optimum solution. Therefore, the general dynamic programming solution procedure has to be modified in order to handle the problem at hand.

The proposed solution is divided into two stages.

The first stage is from the bottom up, that is, from the last operation up to the first. It will proceed operation by operation, determining the optimum path (resource selection) for each operation independently of the previous operation. However, at each operation a review of all previous optimum decisions is made in order to examine the effect of the sequence of operations. The sequence that results in a total path optimum is selected.

The second stage is from top down, which is from the first operation down to the last. It reviews the optimum achieved by examining the effect of the sequence of operations from any operation up to the first operation. The sequence that results in a total path optimum will be used.

The following describes how this technique is applied. For convenience, two auxiliary matrices are constructed (or the same matrix is used with additional values in each box).

In this example the total solution uses three resources (j)—operation (i) matrices:

1. Element value $T_{i,j}$ (it may be time or cost).
2. Total downward value $C_{i,j}$ (i not 1).
3. Path pointers $P_{i,j}$.

The computation starts with operation I-1 and resource 1.

The problem is as follows: to which resource should we proceed from this point in order to arrive at a minimum value? Since this is the last operation, total

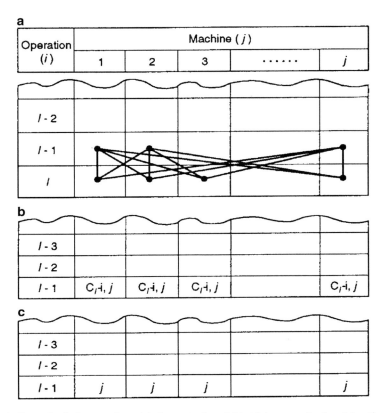

Fig. 2.6 General solution matrices: (**a**) element value; (**b**) total downward value; (**c**) path pointer

downward value $C_{i,j} = 0$, and only the transfer time R should be added when the resource changes. Thus the alternatives are:

$$S_1 = T_{I-1,1} + T_{I,1} (= C_{I,,1})$$
$$S_2 = T_{I-1,1} + T_{I,2} + R_{1,2}$$
$$S_3 = T_{I-1,1} + T_{I,3} + R_{1,3}$$
$$....$$
$$....$$
$$S_J = T_{I-1,1} + T_{I,J} + R_{1,J}.$$

The chosen path will be where S_j is the minimum value. This minimum value is placed in the total matrix as $C_{I-1,1}$. The path matrix lists the resource number of operation I that results in the above minimum value. Thus $P_{I-1,1} = K$. These values are shown in Fig. 2.6.

This process is repeated for operation *I-1* and resource 2 and the resulting values are placed in $C_{I-1,2}$ and $P_{I-1,2}$ and so on until resource J and all values of $C_{I-1,j}$, and $P_{I-1,j}$ are computed. Covering all junction points of operation *I-1*, the solution proceeds upward to handle all junction points of operation I-2, I-3 and so on until the first operation.

The general junction alternatives to be evaluated can be expressed as

$$S_j = T_{i,j} + C_{i+1,k}| + R_{j,k}|_{k=1}^{k=j}$$

The junction to be evaluated is operation i on resource j. Its time or cost is $T_{i,j}$. From this point it is possible to proceed downward to perform operation $i+1$ with one of the available resources. The optimum solution for each resource in operation $i+1$ is the total $C_{i+1,k}$ and is independent of the path by which it was reached.

The term $R_{j,k}$ is the transfer time covering the expenses caused by shifting the work from resource j to resource k. The value of the transfer time is path-dependent. It is possible that by changing the sequence of operations, no transfer time should be added. This case occurs either when the path from the current operation down to the last operation passes through the current resource and the operation that uses that resource can be shifted upward, i.e. to be performed right after the current operation, or when the current operation can be shifted downward to be performed right before the other operations. In such cases, transfer time has already been added to the total and no extra time is required.

The information on whether an operation can be shifted upward or downward is made available by the priority code. Figure 2.7 demonstrates such a case (upward stage).

4.2.1 Upward Stage

The optimum path from operation 10, resource 6 is (i,j) 10,6; 11,7; 12,4; 13,4, as shown by the heavy line in the figure. Transfer time has been added twice so far. The junction of operation 9 resource 4 is evaluated. One of the alternatives is to proceed to operation 10 resource 6. This calls for adding

$$T_{9,4} + C_{10,6} + R_{4,6.}$$

Scanning down the path reveals that resource 4 is selected for operations 12 and 13. If the sequence of operations can be altered to read 9, 12, 13, 10, 11 (as shown in part 2 on Fig. 2.7), only two resource transfer times occur in this path and no extra transfer time should be added. Thus, the total value should be $T_{9,4} + C_{10,6}$. If this is the best alternative for the junction of operation 9 resource 4, the operation sequence will be altered.

This computation and the path checking are performed for any alternative. Thus, when evaluating the alternative of proceeding from operation 9 on resource 4 to

4 Stage 3: Routing Generator

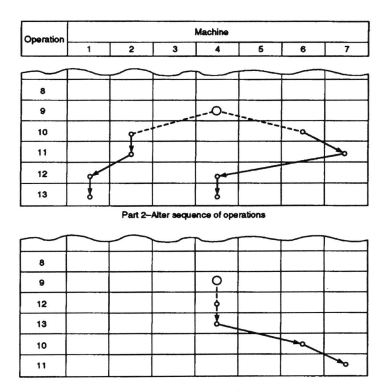

Fig. 2.7 Selection of the sequence of operations—upward stage

operation 10 on resource 2, transfer time must be added (Fig. 2.7), since resource 4 does not participate in the above path.

The priority number indicates whether the sequence of operations can or cannot be changed.

The value in the first row of the total matrix ($C_{1,j}$) represents the total cost or time to produce the part when starting with any one of the available resources (j). The resource chosen for the first operation is the one with the minimum value of $C_{1,j}$. The path matrix will then lead through the resource selected for the other operations and to the sequence of operations.

4.2.2 Downward Stage

The above solution involved changing the sequence of operations by looking downward and saving transfer time. It could not predict the resource selection of the upper part of the matrix. To improve the solution, a second stage of computation is used. In this stage, the operations are examined from the first operation down to the last one on the computed path to check whether a change in sequence of operations will reduce total machining time. Figure 2.8 demonstrates this case.

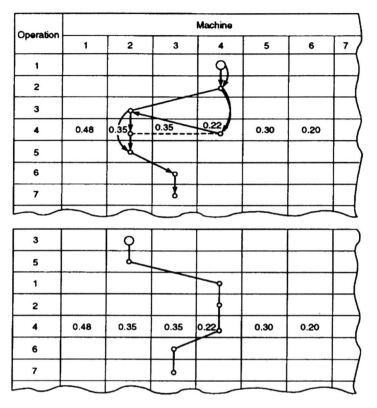

Fig. 2.8 Selection of sequence of operations—downward stage

Scanning the total value of operation 1 indicates that resource 4 results in the minimum value. Thus, resource 4 is selected for operation 1. The path matrix leads to resource selection for the other operations. This path is shown by the line in Fig. 2.8.

Operation 4 has a lower value when performed on resource 4 than when performed on resource 2: $T_{4,4} < T_{4,2}$. However, it was not selected because

$$T_{3,2} + T_{4,2} < T_{3,2} + T_{4,4} + R_{2,4.}$$

The transfer time $R_{2,4}$ must be added since resource 4 is not available on the lower side of the path. Looking from top down, we know that resource 4 is available, and if according to priority code operation 4 can be moved forward, no transfer time should be added. Examining junction 3,2 results in

$$T_{3,2} + T_{4,2} > T_{3,2} + T_{4,4} + 0.$$

4 Stage 3: Routing Generator

Table 2.7 Complete parameters on the matrix

Op.	Pr.	Rel.	User	R#1	R#j	R#J	Min. cost
010	0	0	0	2.28		2.36	1.62
020	010	6	0	1.08		1.18	0.96
030	020	0	0	1.64		1.12	1.12
040	030	0	0	7.96		4.28	3.57
050	010	0	0	1.36		1.48	1.17
060	050	2	0	17.04		8.82	6.75
070	020	0	0	0.52		0.56	0.52
080	070	0	0	1.28		0.94	0.88
090	080	0	0	1.20		0.90	0.86
Total				34.36		21.64	17.45

Thus the sequence of operations should be modified to read: 1,2,4,3,5,6,7,... as shown in Fig. 2.8.

If operation 4 cannot be processed before operation 3, this change of sequence is not allowed. However, it might be possible to process operations 3 and 5 prior to operations 1 and 2.

This means that operation 4 is not going to be pulled up, but rather that operations 1 and 2 are going to be pulled down. This will result in a resource selection and sequence of operations as shown in the bottom part of Fig. 2.8.

4.2.3 Complementary Stage

In solving the matrix for a search for process planning—resource selection and sequence of operations, three constraints are imposed, which are: priority code, relationship code and user request.

Priority code: this code indicates the number of operations that must be performed before the present operation. Its affect was discussed in the previous sections.

Relationship code: this code indicates which operations must be processed on the same resource with the same fixture. Usually it is set by the geometric tolerances specified on the part drawing. The only way to meet the geometric tolerances is to keep the part on the fixture while machining all datum surfaces related to the tolerance. By this method, it eliminates any resource errors.

The matrix will keep this constraint while deciding on the sequence of operations. User code—has the same effect as the relationship code, but it is set by user request.

The complete matrix is shown in Table 2.7.

5 Conclusion

The proposed method results in an optimum selection of resources and sequence of operations. The selection and decision process is purely mathematical and is not affected by intuition or rule of thumb as used today.

An example of rule of thumb is to use old, inaccurate resources for rough cuts and accurate resources for finish cuts. Such a decision will, when true (for large quantities) automatically be reached by the proposed method. However, for small quantities, a different decision might result. The decisions and thus the process planning take the following into consideration:

- *Quantity* The quantity affects the transfer time. The larger the quantity, the lower the value of the transfer time. Thus it is more economical to split the process among several resources, each one suited to some of the operations.
- *Resource capability* Resource time and cost are adjusted for each available resource.
- *Machining incapability* This is introduced by inserting high values for machining time and cost. Thus, it is always more advantageous to add transfer time and bypass this resource.
- *Machining type* Machining time is the same in any resource type, depending only on the power, speed and so on. The difference lies in handling time. The NC, DNC or machining centers resources perform such operations as adjusting the tool or changing speed much faster than a universal manual resource. Automat requires a long setup time, but has the ability to do a fast return and change tools in a short period of time. The handling timetable can have many columns, one for each type of resource. Thus, the general matrix solution can handle different resource types and select the most economical resource type for the job.

Chapter 3
Production Module

Abstract Production planning is by nature a very simple task. However, traditional production planning systems and notions make these systems very complex and unproductive as decisions are being made too early in the manufacturing process.

This chapter presents a different approach, one that introduces flexibility to the manufacturing process where routing is a variable and each order is treated as a unit. Therefore bottlenecks cannot be created and disruptions are resolved automatically.

1 Introduction

The task of the production module is to plan and produce products according to management orders and policy. The major conventional objectives of the task are:

- Meeting delivery dates.
- Keeping the capital tied down in production to a minimum.
- Minimizing manufacturing lead time.
- Minimizing idle time of resources.

Theoretically, to meet these objectives is actually a very simple task. The plant receives orders that define the product, the quantity and delivery dates. The resources of the plants are known, the product's bill-of-materials is known. The task of production is to make sure that the orders will be ready on time, that's all.

It does not seem that it calls for innovation and creativity, as the production processing steps are known and are dictated by series operations that must be performed. Creativity, if needed, is just to solve disruptions such as: a machine fails; a tool breaks; employees are missing; orders change; parts are rejected and reworked; power failures occur; etc. there might be a mismatch between load and available capacity, unrealistic promised delivery dates. Most of the disruptions are a result of the stiffness of a system in which decisions are being made too early in the manufacturing process.

By a different approach, one that will introduce flexibility to the manufacturing process, most of the disruptions will be solved by elimination. Such a system is the subject of this book.

1.1 Traditional Approach

The traditional approach to design of manufacturing systems is a hierarchical approach. The design is based on a top-down process and strictly defines the system modules and their functionality. Communication between modules is strictly defined and limited in such a way that modules communicate with their parent and child modules only. In a hierarchical architecture, modules cannot take initiative; therefore, the system is sensitive to perturbations, and its autonomy and reactivity to disturbances are weak.

The objective of each module is clearly defined and optimized by the system. However, the criteria of optimization are not always synchronized with the total objective.

Local optimization of a single operation does not necessarily lead to optimization of the item. Item optimization does not necessarily lead to optimization of the product. Product optimization does not necessarily lead to optimization of the product mix, and product mix does not necessarily lead to optimization of the business.

Hence, each method solves secondary objectives while ignoring the real task of production scheduling i.e. to make sure that orders will be ready on time. Moreover, the planning and execution regards routing as static and unalterable, therefore it robs the shop of production flexibility and efficiency as it tries to optimize with an unoptimized routing

The basic notions of hierarchical approach techniques are:

- Use the "best" routing for the job.
- Using the "best" routing for maximum production optimization will result in the shortest throughput.
- The larger the batch quantity, the better the productivity.
- Job release to shop floor based on MRP (or ERP) will assure maximum efficiency and meeting delivery dates.

1.1.1 Use the "Best" Routing for the Job

Recent research in the field of process planning (CAPP) highlights the fact that routing is an obscure term. There is no such feature as the "best" process. For each task there might be many different routings, which differ from one other by processing time and processing cost and objective. Usually the shorter the processing time, the higher the processing cost (Table 3.1).

1 Introduction

Table 3.1 Shows 22 alternative process plans for item "CROSS"

Alternative number	Total cost	Total time	Max. time
1	23.76	5.94	5.94
2	19.02	6.34	6.34
3	15.54	11.10	11.10
4	24.98	12.49	12.49
5	19.80	5.84	4.48
6	17.70	8.94	5.72
7	22.80	7.10	5.00
8	19.57	6.59	5.89
9	16.97	6.22	4.83
10	15.53	9.19	5.72
11	18.15	7.45	5.35
12	15.18	9.99	7.73
13	14.98	12.42	6.40
14	15.19	10.47	9.08
15	24.90	13.50	11.40
16	16.09	13.34	9.99
17	14.94	12.98	6.40
18	17.66	8.90	3.88
19	14.90	10.09	6.23
20	14.34	10.81	5.72
21	14.83	9.36	6.85
22	14.30	10.77	3.88

Alternative 1—provide max production
Alternative 22—provide min cost
Alternative 9—provide max profit
Alternative 3—provide min investment
Alternative 13—provide years for ROI
Alternative 8—provide quantity for ROI
Which one of the routings is "best" for production and management planning???

1.1.2 Using "Best" Routing for Maximum Production

Using the "best" routing for maximum production optimization will result in the shortest throughput.

The term "optimized routing" is somewhat misleading. A routine based on criteria of optimization of maximum production actually means that the time to produce the item, as a standalone, is minimum. However, it does not refer to the elapsed time to produce a product or a product mix. In many cases the maximum production routine will result in the longest processing time of a product mix.

In process planning there are several levels of optimization:

- Optimization of a single operation.
- Optimization of an individual item.
- Optimization of producing a product (several items).

- Optimization of producing a product mix.
- Optimization of factory business.

Each one has different requirements and must consider the elements of production differently and probably will result in a different recommended routine.

1.1.3 The Larger the Batch Quantity, the Better Productivity

This notion seems logical and reasonable. The larger the batch sizes the lower the setup cost and time consumed per single item.

> However, as K.R. Baker in his book *Introduction to sequencing and scheduling* states, "what might at first seem to be undue emphasis on the turnaround criteria is not really so restrictive, in the light of this relationship between flow time and inventory, because mean flow time actually encompasses a broader range of scheduling-related costs". Furthermore, his theorem states that "Mean lateness is minimized by SPT (Shortest Processing Time) sequencing".

SPT is affected by order quantity and routing. Increasing quantity increases processing time, and thus affects the throughput. The conclusion might recommend not increasing the quantity by combining items from different orders. This conclusion lacks clarity and specificity and has no academic foundation.

In order to check these hypotheses, simulation techniques were used. In this simulation three orders, each for a quantity of 40 units are considered, each order is composed of 10 items, in four bills-of-material levels. The number of operations per item ranges from 1 to 7, overall there are 60 operations. There are 15 resources that cover all processes required, they are: milling machines, lathes, drill presses, saws, robots, and mechanical assembly.

The simulation uses five modes of scheduling. One mode combines all orders to be produced as one batch of 120 units. Then to use two patches of 60 units each, and a third combination to produce each order separately. Another simulation examines the case in which the batch sizes are divided into two batches, one of 80 units and the other of 40 units. In this case it is tested if a sequence of loading non-symmetric batches has an effect on processing time.

The results are shown in Table 3.2 and refers to the number of periods (period = 2 h) that it takes to process the relevant orders.

The simulation clearly shows that regardless of the criteria of optimization used, the number of periods to produce the same quantity diminishes as the batch size is reduced.

1.1.4 Scheduling with Limited Flexibility

In addition to strict criteria of optimization of minimum cost and maximum production, we tested the effect of introducing a limited degree of flexibility by mixing these two.

1 Introduction

Table 3.2 Throughput versus lot size

	Criteria of optimization			
Batch quantity	Max. prod.	Min. cost	Cost/prod.	Prod./cost
120	49	53	55	59
60 + 60	46	42	36	40
80 + 40	44	46	35	46
40 + 80	46	45	41	44
40 + 40 + 40	45	37	29	36

Table 3.3 Throughput versus flexible production planning

		Criteria of optimization			
Batch quantity	No. of idle periods	Max. prod.	Min. cost	Cost/prod.	Prod./cost
40 + 40 + 40	Never	45	37	29	36
40 + 40 + 40	10	32	37	29	27
40 + 40 + 40	3	25	20	19	22
60 + 60	Never	46	42	36	40
60 + 60	10	32	38	29	31
60 + 60	3	26	32	27	28
40 + 80	Never	46	45	41	44
40 + 80	10	33	40	37	42
40 + 80	3	28	36	29	29
80 + 40	Never	44	46	35	46
80 + 40	10	39	46	35	46
80 + 40	3	28	36	29	29
120	Never	49	53	55	59
120	10	42	53	39	43
120	3	31	42	34	37

Cost/Production criteria means: starts scheduling with the minimum cost routing, but if an operation has to wait for a resource for a limited number of periods, switches to operation of the maximum production routing.

The second case, designated as Production/Cost, means to start scheduling with the maximum production routing, and if an operation has to wait for a resource, then switches to operation of minimum cost routing.

The simulation results are shown in Table 3.3. It shows that scheduling with limited flexibility within the Cost/Production algorithm improves the throughput time. Furthermore, it shows that the use of Cost/Production method is superior (in this simulation) to the Production/Cost scheduling method, and that treating each order as a separate unit reduces the throughput time.

The indication that limited flexibility improves the throughput time of producing a product mix, led to investigate the effect of flexible dynamic scheduling.

1.1.5 Flexible Dynamic Production Planning

Table 3.1 shows that an item may be processed by many routings. A conventional method selects a single routing, and uses it for production planning and scheduling. The last section showed that by using two routings the throughput could be improved. But why stop at two routings and not all possible routings? The simulation program does it.

The simulation program anticipates a resource overload (bottleneck) and attempts to resolve it in the planning stage. The logic is that whenever an operation is *"free"* to be processed but the "best" resource for that operation is occupied, the system checks if it is economically effective to assign that operation to another resource. A *"free"* operation is defined as an operation that at that instant is available for loading. The "Best" resource is defined as the resource that, by the selected criteria of optimization, results in minimum (cost or time).

The following simulation tested this algorithm. A change of resource and routing in case of overload is done in three cases: Case one—never change (this case is presented just for comparison of results). Case two—change a routing in case that a free operation must wait for 10 periods. Case three—a free operation has to wait for three periods. The results are shown in Table 3.3.

The results of this simulation confirm the previous conclusions regarding the choice of treating each order individually, and in addition it clearly shows that the increase of flexibility decreases significantly the number of periods required to produce the product mix (which was tested by the simulation).

1.1.6 Summary

The traditional approach to planning and execution regards routing as static and unalterable, therefore the planning is simple, but it robs the shop of production flexibility and efficiency.

The results of introducing flexibility to the system by regarding routing as a variable leads to the following conclusions:

1. Each order, with its product structure, should be treated individually. Do not attempt to increase quantity by combining similar items into one processing batch. In one case, processing time increased by 43% when items were combined (from 37 to 53 periods).
2. Selecting a routine based on maximum production criteria of optimization does not assure reduction of processing time.
3. Limited flexibility reduced the processing time from 53 periods to 22 periods.
4. Total flexibility reduced the processing time further to 18.16 periods.

2 Production Management Strategy: Roadmap Manufacturing

The proposed manufacturing strategy approach makes use of the following notions:

- There are infinite ways of meeting design objectives.
- In any design about 75% of the dimensions (geometric shape) are nonfunctional (fillers). These dimensions can vary considerably without affecting the design performance.
- There are infinite ways of producing a product.
- The cost and lead time required to produce a component are functions of the process used.
- Transfer of knowledge between disciplines working to produce a product should not be by transferring decisions, but rather by transferring alternatives, ideas, options considered, reasoning etc.
- The company database should be "open" and available to all disciplines.
- Any decision made is subject to change.
- Regard routing as a variable.

2.1 Roadmap Notions

Roadmap notions are what practically any of us does in personal life.

For example: If you plan to go from point A to point B, than you study a map and plan the optimum route to take. This is a present time decision. However, at another time when you have to travel the same way, say at night, you might change the route; in winter you probably will look for a route with maximum shelter from precipitation. In summer you might choose a route that protects you from the sun. In springtime you might choose a route with a nice view. Despite the original routing decision, if you run into disruptions, such as a blocked road (bottleneck), red traffic light or with crowded traffic you might decide that instead of waiting in line, it is better to consult the road map and change the route in order to find a path with no obstacles. Such a change is decided at each junction. It might be a longer route but it will be faster in time. The original decision must not prevent one from adapting the new route.

Similar strategies can be applied to production management. The manufacturing system we are presenting proposes to supply each manager with a "road map" that is stored in the company database, and allow him or her to deviate from the original plan while accomplishing the production program and target objectives assigned. The proposed method will introduce flexibility and dynamics, thereby increasing company efficiency and customer satisfaction.

The problem is that, in spite of present day technology, most manufacturing databases include only decisions, without revealing why and how they were arrived at, and therefore the "road map" indicates only one path, or sometimes an alternative; but a single alternative is not a road map. Naturally whoever made the decision considered many options, alternatives, and optimization methods. Regrettably, all such considerations are lost, and only the decisions themselves are transferred and stored in the company database. What should be done is to capture all the relevant data leading up to decisions and store it for future reference.

For example, the task of the process planner is to select out of the tremendous number of alternatives, the most economical process according to his or her decision. The planner naturally does what seems to be best. However there are several criteria of optimization such as: maximum production; minimum cost; maximum profit. Each one of them will result in a different routine. However, the process planner is usually neither an economist nor a production planner; therefore the planner should not make decisions that are beyond his or her field of expertise. Furthermore, there are several criteria of optimizations that are affected by routing such as:

- Optimization of a single operation.
- Optimization of an individual item.
- Optimization of producing a product.
- Optimization of producing a product mix.
- Optimization of factory business.

Single routing in a company database cannot accommodate all these criteria; only a roadmap method is adequate to the task.

2.2 Production Planning

The objective of this stage is to plan activities in a manner that order delivery dates will be met. The outcome of this planning is the order release to the shop floor for execution. The order release must be practical otherwise the shop floor will not be able to follow the given plan. If unrealistic job orders are released to the shop floor, one cannot expect that all jobs will be finished on time. From the released list of jobs the foreman will select, on the basis of the available employees and their expertise, which job to execute. The foreman's decision criteria for job execution do not always coincide with those of the planning department.

To arrive at a practical plan and job release, the planning must be the same, or at least very similar to the scheduling of jobs on the shop floor. It must be able to show the shop floor personnel the exact schedule, and confirm that all the released jobs may be done at the planning period. The shop floor may use any scheduling it desires, but it must complete all released orders on time.

Thus the plan must consider: the available resources, i.e. a plan with finite capacity; the product structure—i.e. when one item is delayed, allowing a shift in

2 Production Management Strategy: Roadmap Manufacturing

the planning of all other dependent items to the time that it is really needed; the degree of flexibility available in selecting the routing, i.e. considering the routing as a variable.

The roadmap method of planning assumes finite capacity and availability of the product tree. The detailed capacity plan is transformed into a real schedule. The schedule is the basis for dispatching jobs to the shop floor and determining how and when the released product mix has to be produced. However, the shop floor is still free to produce the product mix by any other routing that facilitates the solution of problems caused by disruption, with the restriction that the released product mix, as specified for the period, must be completed on time.

The planning steps are as follows:

1. Determine stock allocation priorities.
2. Make stock allocations.
3. Adjust quantities by economic considerations.
4. Check capacity planning—machine loading.
5. Release jobs for execution.
6. Establish shop floor control.

2.3 Stock Allocation Priority

A working plant is a dynamic environment, subject to many changes and unplanned interruptions, which may lead to the accumulation of un-required stock; these changes and interruptions might include:

- Interruptions in the shop causing early or late finish of jobs.
- Reject rate being higher or lower than anticipated.
- Excess units of the other items are left over after assembly.
- Customer orders being added or deleted.
- Quantities and delivery dates being altered.
- Purchasing restricted by package size.
- Economic consideration of lot size.

All these factors lead to the accumulation of stock. This stock can often be utilized later in further manufacturing. The objective is to plan the activities to be performed in order to meet the goals of the master production schedule, while accumulated stock is taken into account.

The objective of this step is to *set priorities* for stock allocation, i.e. to which order and item to allocate the available stock.

The strategy is to allocate the stock according to the critical order, where "critical" entails determining the earliest time at which the lowest level item must start its processing. The earliest time might be in the past, or on some future date. To meet this strategy the first step is to build the product structure on a time element scale, instead of on a level-base, as shown in Fig. 3.1.

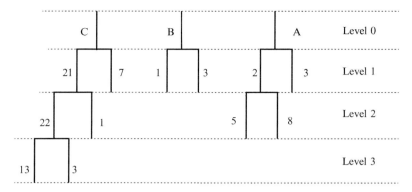

Fig. 3.1 Level-based product structure

This figure shows three products A, B, C and the items that are in each product at each level. Level 0 is the product (or order); the lowest level in this figure is on order C and is level 3, which is also referred to as "low level" and items 3 & 13 are referred to as low level items. The connecting lines represent the relationship of the product, its sub-assemblies and the items and do not represent the time to process an item.

In order to determine which order is the critical order, the level-based product structure is converted into a time-based product structure. The name of the order is retrieved from the level-based product structure (level 0) and the matrix is called to generate a process base for the order quantity for that item.

2.3.1 Time-Based Product Structure

The conversion of a level-based to a time-based product structure starts with order delivery date down to lower levels. Call the matrix to compute the assembly time of the product (level 0), multiply by the quantity, then convert the sum to delivery date units. This value is subtracted from the delivery date. The result is the starting point in processing time of the level 1 items. Again the time for processing each one of the items in level 1 of that order and its quantity is reduced from the previous level end point.

The processing time of such a process is given for each single item. The total processing time is computed by multiplying the quantity by the processing time of a single product. Convert the computed total length of time to the time scale (let us say days) and subtract it from the delivery date of the order. Draw a line starting from the order delivery date backward, at the computed length. The end point of this line indicates the date at which the assembly (processing of the order) must start in order to meet the delivery date. Record this line on the time-based product structure.

Next, treat one by one all items of level 1 of the same order, regarding the start processing date of level zero as the delivery date of each item of level 1. Call the matrix to generate the economic process, compute the total time and convert it to the scale time and draw the connecting line by this length. Repeat this process to all levels of the order, and to all orders on the file.

Example

Figure 3.1 shows three orders for products A, B, & C. The computation may start with any order. For example suppose it starts with order A. The assembly of order A is treated first, and the matrix issue the time to assemble. This time is indicated by the length of the line, from the delivery date of order A backward in time.

The next step for the product A structure is sub-assembly 2. The quantity is computed by the order quantity multiplied by the number of sub-assembly 2 in one product A. The system turns to the matrix and retrieves the time to assemble sub-assembly 2. The assembly ends at the beginning of assembly A and starts at the due date minus the assembly of sub-assembly 2.

Next, the sub-assembly 2 is composed of items 5 and 8. Their due date is the beginning of sub-assembly 2. The quantity of item 5 is computed, by multiplying the quantity of item 2 by the number of items 5 in assembly 2. The system turns to the matrix and retrieves the time to produce item 5. The processing due date is at the beginning of sub-assembly 2 and ends at this date minus the processing time of item 5. A similar process is initiated for each item.

The due date for item 3 is the beginning date of order A. Again the matrix supplies the processing duration.

Note: the quantities of each item will allow for a scrap factor.

Figure 3.2 shows the time-based product structure of three orders each with a different delivery date.

The level-based product structure is regarded as a master structure, and refers to all company products, while the time-based product structure is a working structure and refers only to the open orders of the company. The working product structure represents the activities that should be taken in order to fulfill the customer orders.

2.4 Stock Allocation Method

The objective of this step is to allocate the available items in inventory to the available orders and adjust the early starting dates of each item accordingly. It is done by scanning all the orders and finding the critical item. A critical order is defined as the order that its lowest level item has as the earliest starting date for all orders, (i.e. the item with the earliest starting date.).

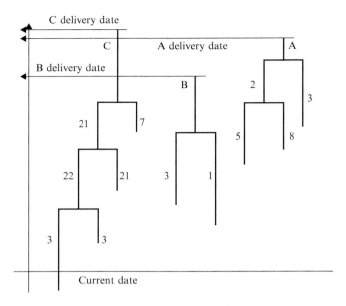

Fig. 3.2 Time-based product tree

When the item with the earliest starting date is determined, the product structure for this item is examined in order to find the product and the order (i.e. the level 0) (see Fig. 3.3). The inventory is checked to see if this level 0 product is available in stock. If available, it is allocated to this order. The quantity of this order is reduced by the available quantity, and the product tree is rebuilt with new quantities per item and new starting dates. In this case the starting time of the lowest level item will be changed and another order might become the critical one. The procedure is repeated by scanning all orders to find the "new" critical item. If the ordered product is not in stock, the availability of the next level item using examined using the same procedure.

For example : examining Fig. 3.3 reveals that item 3 of product C has the earliest starting date; therefore it is regarded as the critical item. However, the allocation priority should be given to level 0 item of the critical item, since there will be no need for that item if higher level items are available in stock. The chain of the critical path is: items 3 – 22 – 21 – C. Therefore, the system checks the inventory for availability of product C.

In case that item C is not in stock, or in partial quantity, such quantity is allocated to item C and the quantities of all items in the tree are adjusted accordingly and a modified time-base is constructed. The allocation proceeds, as described, to items on the lower level.

Suppose that item C is available in stock for the entire quantity; then all product structure of C is marked as available, and is erased from the time-based product tree. In this example, only products A and B are considered.

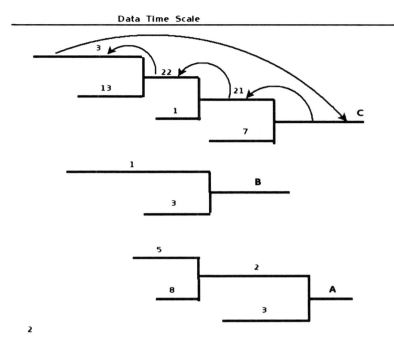

Fig. 3.3 Stock allocation sequence of priorities

After each allocation a check is made to find the current critical item. In this example item 1 is the new critical item. The path to the low level item is: 1—B. The system checks to determine if item B is in stock; if it is, it is allocated. If item B is not available in stock, a check is made to determine if item 1 is available; if it is, it is allocated to item 1 of product B. In case of partial quantity, the remaining quantity is reduced, and thus the time to produce item 1 is reduced. Item 1 of product B is marked as "treated" to make sure that it is not considered again as the critical item. Such an intermittent state is shown in Fig. 3.4. In the next step, examining the time product structure indicates that item 5 of product A is the critical one The process continues till all low level items are marked as treated.

This method assures that allocation does not consider the delivery date of an order but instead makes sure that the critical items get a priority. This point is illustrated by the time structure shown in Fig. 3.3. Item 3 appeared in orders for product A B & C. The early delivery date is for product B, next to product A and then for product C. However, the critical sequence of allocation should be, according to the state indicated in Fig. 3.3, product C then B then A.

Note: The procedure described above calls for scanning the bill-of-materials many times, both up and down directions, (i.e. from level 0 to the low level and vice versa). One might assume that this would take a lot of computer time and therefore would not be practical. That would be true if the file management of the product structure were constructed in a conventional method. However, if the

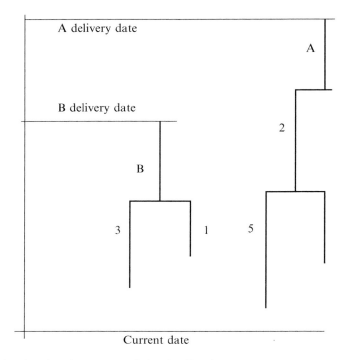

Fig. 3.4 Time-based product structure during the allocation process

product structure is written in a method that supports the needs of the application, it takes just a few seconds to run 23 orders on a PC, including the time to call the matrix and retrieve an optimum process plan for each item on the product tree. A suggested method is to construct the product tree file management as one record for each branch of the tree, from level 0 to the low level item, although each user can choose his or her own file management structure.

2.5 Adjust Quantities

At the end of the stock allocation step, the product structure includes only those items that have to be produced, or purchased. The working product tree is not similar to the master product tree, as some items might be missing altogether, others might have a different quantity.

For each low level item there is a flag that indicates that this item and its branches has been dealt with. The first task in this step is to remove those flags. The product structure status now is the working product structure where all the available items are removed. Work in process is regarded as available items. The quantities of each

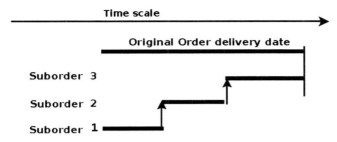

Fig. 3.5 Delivery date for suborders

item are the net requirements. However, the manufacturing batch size is not the net requirement but the economic lot size. Therefore, at this step the economic lot size is computed.

The economic order quantity can be computed by the roadmap method, any other method, or by external constraints. If the order quantity is larger than the EOQ, then a check should be made if it will be more economical to split it into several suborders. The number of suborders should be computed by the following equation:

Number of suborders = (Confirmed order quantity)/(Economic lot size).

The number of suborders should be an integer. Rounding the result to the nearest integer is recommended (i.e. 2.4 becomes 2 and 2.6 becomes 3).

The quantity in each suborder is computed by the following equation:

Manufacturing batch size = (Confirmed order quantity)/(Number of suborders).

The delivery date for the suborders is computed in such a method that the finished ordered products will be continuous, Fig. 3.5 demonstrates the delivery dates in case an order was split into three suborders.

Whenever an order splits to several suborders, the starting dates indicated on the original order are erased, and the matrix is called again to compute, in the same manner as described before, the new starting date of each item in each order. For simplicity of programming, re-compute each suborder as if it is an original order (instead of copying the dates).

2.6 Capacity Planning: Resource Loading

The working product structure lists the items that have to be processed. It is built in a format that provides a connecting link upward to its parent item. Pointing any individual requirements to their specific source is a significant feature; it provides an upward tractability from component to parent item, all the way up to the end item requirement. It is used to:

- Check the source of requirement.
- Trace the effect of component delay on the delivery date of the finished product. For example, if a process operation cannot meet its due date, all forward process operations will be scheduled at the realistic due date of the previous operation. Resource loading considers the order with all its items, and not of any individual item.
- Examine the validity and significance of a system-generated request to change the delivery date of the order.
- Discover the effect of a pending engineering change on a customer's order or trace upward to the product serial number on which the change will become effective.
- Maintain the customer's identity down through lower level component orders.

Resource loading employs a table in which each available resource is represented by a column and each period is represented by a row. The data in each cross section slot of resource and period indicates the state of the resource at that period. If the content of the slot is a data, it indicates that the resource at that period is occupied, processing a specific operation of a specific item. If the slot is blank (empty) it means that the resource at this period is idle.

Resource loading is forward planning. When a job is allocated to the resource, the appropriate position(s) in the column hold(s) the order and the item code. Empty positions indicate idle resources.

A critical planning path is defined as the path starting at the lowest level item in the working product structure with the earliest starting date, through its sub-assemblies up to the product. The critical path will have priority in resource loading. However, all the items for the sub-assembly have to be available before the assembly can start, i.e. all independent items that go into the critical sub-assembly will have a priority greater than the sub-assembly.

The priority of such items is related to their starting date. By this method the critical product has loading priority. The items in the critical path of the critical product have loading priority. This method *resolves the problem of competition of jobs over resources*. It guarantees the loading will be assigned to the most desirable job. In case that a job requires a time slot at a certain period, and the resource is occupied by another job, it means that the other job had priority over the present one, since it was treated earlier.

If the required resource will become available in a short period of time, then the job in the queue will wait. However if the number of periods the job has to wait in the queue is large, than the system will call the matrix in search for an alternate process. An economic model will be used to determine if it is more economic to wait and delay the operation, or to use another resource (process).

The routing is regarded as a variable, and is generated at time of need. The matrix features of forced process planning, (which forces the matrix solution to use an indicated resource, in this case a resource that is idle at the required period) is used.

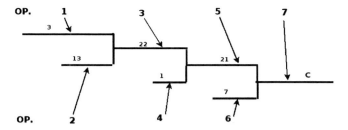

Fig. 3.6 Sequence of resource loading

In many cases the difference in time between the "best" process and the first (and second) alternate process is negligible. By this method *no bottlenecks* in production will result, and no bottleneck resolving procedures are needed.

2.6.1 The Loading Procedure is as Follows

Scan all the low level items of the working time-based product structure and find the item with the earliest starting date. Run over the product structure to find the product and the order, as shown in Fig. 3.3. This order has priority in resource loading. The loading is done from the lowest level item, from its first machining operation and forward. Subsequently, call the matrix with the item name and quantity and retrieve a process. The retrieved process indicates the number of resources, the name of each resource and the time for the operation (including setup and penalties).

Start with the first operation, multiply its time by the quantity (the real quantity and not the bill-of-material quantity) and divide by the period scale of the row. Determine how many periods are needed for this operation. Run over the column of the appropriate resource and search for an idle period. The search starts at the operation's early start period (depends on the previous operations and the product tree). When the idle time is found, the name of the item is inserted in the row. In the case that the early available idle period is too far ahead of the early start period, the matrix will be called to generate an alternate process. The alternate process will attempt to reduce the waiting time by employing a different resource that is idle at the required periods and that it is economical to change. The initial proposed process plan matrix solution is based on the maximum profit criteria of optimization. However, in many cases the difference between the alternate process plans is negligible. Blocking the occupied resource (using the "machine blocking" feature) and resolving the matrix, will generate the next alternate routing. If a known machine is idle, the "forced process plan" feature will be used and the economics of using that process plan are examined. This process may be continued until the available space is found.

The sequence of items to be loaded is shown in Fig. 3.6. Assume that item 3 of product C has the earliest starting time. Therefore product C with all its items is the critical path and has priority. The critical path is items 3 22 21 C. Therefore item 3 is the first one to be loaded. The matrix for item 3 is called and the process plan is retrieved. All its operations are loaded. Next on the critical path is item 22. However, as it is a sub-assembly, it cannot be assembled without processing item 13. Therefore item 13 will be treated and loaded next, before item 22. Next on the critical path is sub-assembly 21. However, it cannot be assembled without processing item 1. Therefore item 1 will be treated and loaded next. Next on the critical path is assembly C. However, it cannot be assembled without processing item 7. Therefore item 7 will be treated and loaded next. The sequence of loading will therefore be: 3 13 22 1 21 7 C. Note that item 13; 1; and 7 are independent items. Their starting period might be the first period, the same as item 3. If there is competition over resources between these items, the sequence will automatically assign the resource according to the product priorities. If the slack between the starting of sub-assembly 22 and the end of processing item 13 is too large, the starting date of item 13 will be delayed. In such cases, (including the delivery date of the product) safety periods will determine the finish period for item 13 and it will be loaded backward from this period. Initially the loading is done on a temporary basis. The loading is marked on an auxiliary resource period table that supplements the actual table. This temporary loading checks if the delivery date of the order is being met, and what is the slack if any. In case that the delivery date is not met, the temporary loading is erased, and a new attempt is made, this time using the maximum production criteria of optimization to generate a process from the matrix.

At the end of loading product C it will be marked as loaded. The process will repeat itself for the remaining orders on the working product structure, till all its products will be marked as loaded.

Carrying out the planning actions as described above, results in:

- Minimum processing lead time.
- Meeting delivery date.
- Resource utilization.
- Minimum work in process.
- Minimum capital tie down in production.
- Elimination of bottlenecks in production.

2.7 Job Release for Execution

Capacity planning plans the jobs to be carried out in the plant in order to meet customer orders. It is a long range planning. The actual processing is done on the shop floor, which requires a short-term production plan. The objective of capacity planning is to determine what job to release to the shop floor for execution. Different companies may have a different strategy regarding the length of the period for which the jobs are released to the shop floor. Too long a period might result in exceeding the delivery dates, while a too short period might cause idle time at shop floor level. A period of 1–6 days is commonly used.

Capacity planning is a very important planning task, bridging the gap between long and short term planning. To carry out the production plan, jobs have to be released to the shop floor for execution. The jobs in the capacity plan that appear on several early periods are to be released to the shop floor for execution.

Before job execution can start, some auxiliary jobs have to be performed. The auxiliary jobs are:

- Fixture design and building.
- Tool preparation.
- NC program generation.
- Material preparation (inventory management and control).
- Material handling (transport).
- Quality control (preparation of method and tools).
- Setup instructions and setup.
- Job instruction.

Therefore resources for performing these auxiliary jobs have to be released for production, at the appropriate department. The jobs for a medium range period, whose length depends on the specific factory procedures, are used to alert the auxiliary job departments of the jobs that are going to be released for production.

One can distinguish two main strategies in dispatching auxiliary jobs. The first strategy, which is to be preferred when execution of an auxiliary job takes a relatively long time, deals with an early release during capacity planning. In this case the main job is released for execution only after the auxiliary job has been reported ready. The second strategy, which is preferred when the execution of the auxiliary job takes a relatively short time, is to release these jobs at the beginning of the scheduling period.

3 Shop Floor Control

Shop floor control (SFC) in a batch-type manufacturing environment is regarded by the current research community as a very complex task.

It claims that the complexity is a result of the system approach in which inflexible decisions are being made at too early a stage in the manufacturing process. It proposes a method that introduces flexibility and dynamics and thus simplifies the decision making in production planning. The SFC method, which is a module of a production management system, proposes that in order to introduce flexibility, routings should be regarded as a variable. Each expert will generate routines that meet needs at appropriate times, and will thereby increase dramatically manufacturing efficiency.

Table 3.4 Process plan matrix

OP.	PR	R1	R2	R3	R4	R5	R6
10	00	3.12(*)	3.17	3.68	99	4.02	3.27
20	00	1.15	1.2	1.71	99	2.05	1.3
30	20	1.49	1.53	2.05	99	99	1.64
40	10	1.30	1.35	1.86	1.86	2.2	1.45
50	40	1.28	1.33	1.84	99	2.18	1.43
60	50	1.51	1.56	2.07	99	99	1.66

Time to process each operation on available resource
PR—The priority of sequencing the operations. The number is the operation that can be processed only after processing operation no. has finished
00—Operation that may be processed at any time
99—Resource that is unable to process this operation

3.1 Concept and Terminology

The proposed shop floor control approach is based on the concept that whenever a resource is free, it searches for a free operation to perform. A *free resource* is defined as a resource that has just finished an operation and the part has been removed, or is idle and can be loaded at any instant. A *free operation* is defined as an operation that can be loaded for processing at any instant. An example would be the first operation on an item for which the raw material and all the auxiliary jobs are available, and is within reach of the resource operator. An intermittent operation is one for which the previous operation has been completed and the part has been unloaded from the resource that performed the previous operation, and is within reach of the required resource.

Example: Process planning of an item is given in Table 3.4. Operations 10 and 20 have a priority of 00; they are both free operations and can be loaded whenever a resource is available. When operation 10 is done, then operation 40 becomes free. When operation 20 is done then operation 30 becomes free. When operation 30 is done, no operation becomes free. Therefore the sequence of operation may be: 20; 30; 10; 40; 50; 60 OR 10; 20; 30; 40; 50; 60; OR 10; 40; 50; 60; 20; 30 and several other combinations as indicated. This shows the flexibility of the system.

The term *operation* has different meanings in production management and scheduling and in technology. **Production management operation** considers an operation as a set of all the activities done on one resource, from the loading till unloading. It does not give any indication of what are the operations. Production management operations (routing) are used for production planning and scheduling, while the technological operations are used for resource setup, and preparing work instructions. **Technological operation** is an individual processing operation (the six operations in Table 3.4). The term open operation in the proposed shop floor control approach refers to technological operation.

The scheduling cycle starts by scanning all resources in search of a free resource. The free resource scans all free operations and lists them. The best operation for a resource can be based on performance objective, such as minimum processing time or cost. This scanning results in a list of candidates for scheduling.

If the list contains only one entry, than that operation is loaded on that resource.

If the list contains more than one entry, then the system allocates the operation with the biggest time gap of processing it on another resource.

If the list is empty, this means that there is no free operation available for processing on that resource. Hence the resource becomes idle, waiting for an appropriate operation. Idleness is a waste of time and such time may be used to process a free operation. Despite increasing processing time, it might be economical. Therefore the system searches for a free operation that the idle resource can perform although not being the best resource for the job, but is economical. One method to compute the economics of using an alternate resource is to compute the difference in time between the "best" and the alternate operation and compare it to the time that the free resource will otherwise be idle.

As an example: suppose that the quantity is 100 units, the best processing time is 5 min. The alternate resource processing time is 6 min and the waiting time is 150 min. Then the economic consideration is as follows:

1. To produce the operation with the best resource it will take $5 \times 100 = 500$ min.
2. To produce the operation with the alternate resource it will take $6 \times 100 = 600$ min, out of which 150 of them are replacing the waiting time.

Therefore the actual processing time is $600 - 150 = 450$. Hence using the alternate resource and working "inefficiently" will save $500 - 450 = 50$ min of elapsed time.

If this next operation is more economical or better in terms of performance for this resource then the following operation is allocated to that resource. Economical or better performance means that this resource is the best for this operation, or that its processing time (or cost) minus a transfer penalty is equivalent to or lower than the best time of that operation.

Transfer penalty is defined as the time/cost to transfer a job from one resource to another. It includes setup time, inspection, storage, material handling, etc.

In case of resource breakdown, no special treatment is needed. It will be marked as busy, hence in no scanning cycle will it be regarded as a free resource.

In case of an item being rejected, the product structure is consulted to determine if it will hold assembly. If so all items required for that assembly are not needed and will be removed from the list of released jobs for the period.

3.2 Algorithm and Terminology

Shop floor control starts with a list of jobs that should be processes in the relevant period. Such a list may be compiled from the matrix production planning module, or from any other source. The list contains:

1. Job number and name
2. Quantity
3. Sequence priority
4. Bill-of-materials

These jobs are free for execution. However, before job execution can start some auxiliary jobs have to be performed. The auxiliary jobs are:

- Fixture design and building.
- Tool preparation.
- NC program generation.
- Material preparation (inventory management and control).
- Material handling (transport).
- Quality control (preparation of method and tools).
- Setup instructions.
- Setup of job instruction.

Each of the free jobs retrieves from the company database the two-dimensional process plan matrix (as shown in Table 3.4) and constructs a 3D matrix process plan as shown in Table 3.5 (3D: Resources—Operations—Items).

The algorithm is based on the following records:

Resource status file keeps the status of the resource throughout of the scheduling period. The data stored is:

- Resource number
- The loaded item, item and operation
- Quantity
- A link to the bill-of-materials
- Resource counter
- Sequence number of entry in the History file

Resource counter is a counter that indicates the remaining time for processing the item. When loaded it is set by multiplying the quantity by the processing time, as indicated by the 3D matrix, and it is updated at each scan cycle by the elapsed time from the last scanning cycle.

History file keeps track of the actual performance on the shop floor. It keeps the following data:

- Sequence number
- Resource number

3 Shop Floor Control

Table 3.5 3D matrix status when R4 is idle

Op	PR	R1	R2	R3	R4 IDLE	R5	R6	BEST	Δ
	I	T	E	M		#3			
10	X	12.5	9.51	5.15	**99**	4.02	6.54	5	
20	X	5.04	3.93	2.55	**99**	99	2.82	3	
30	X	6.28	4.86	2.98	**2.53**	2.47	3.44	5	
40	*00*	*6.38*	*6.12*	*7.05*	*5.78*	*5.93*	*6.83*	*4*	*1.27*
50	40	8.24	6.33	3.67	**2.96**	2.62	4.42	5	
60	50	5.15	99	4.02	**4.86**	2.98	2.53	6	
	I	T	E	M		#5			
10	X	3.12	3.17	4.02	**3.27**	99	99	1	
20	*00*	*13.9*	*10.3*	*10.8*	*9.95*	*12.5*	*99*	*4*	*3.95*
30	20	4.86	2.98	2.53	**4.86**	2.98	2.53	3	
40	20	6.04	4.68	2.90	**99**	99	3.32	3	
50	40	5.76	4.47	2.8	**99**	99	3.18	3	
	I	T	E	M		#7			
10	X	3.12	3.17	4.02	**3.27**	99	99	1	
20	X	6.15	4.2	8.05	**9.3**	99	99	2	
30	*00*	*8.34*	*8.92*	*7.58*	*7.23*	*8.76*	*8.12*	*4*	*1,69*
40	30	2.06	2.11	2.96	**2.21**	99	99	1	
	I	T	E	M		#9			
10	X	4.6	3.60	2.39	**99**	2.05	2.60	5	
20	X	5.96	4.59	2.87	**99**	99	3.28	3	
30	*00*	*11.5*	*12.8*	*11.9*	*11.4*	*13.1*	*99*	*4*	*1.7*
40	30	99	99	99	**99**	1.45	1.72	5	

- Product, item and operation
- Start time
- Finished time

The objective of the history file is to store data for management and production control reports. It can be used to compare planning to actual performance, to arrive at actual item cost, resource load etc.

The scheduling module is based on a *sequence cycle* loop that examines all resources, listed in the *resource status file*, loads the free resources and updates the resource counter. The sequence cycle loop starts whenever processing an operation is finished. At this point the resource becomes idle and a decision has to be made on the next assignment.

Sequence cycle time is the elapsed time between present time and the previous sequence cycle loop. The time is retrieved from a running clock that starts at the beginning of the scheduling process and advances by the working time.

Shop floor control is based on the concept that whenever a resource is free, it searches for a free operation to process. A free operation is identified by scanning the "PR" column of a 3D process plan matrix. Any operation with PR = 0 is a free operation. A free resource is identified by the resource counter as equal to zero (0).

The sequence cycle loop scans all resources and checks the field resource counter.

If the counter is zero it means that it was idle in the last scanning cycle, and will be treated as such (see next case).

If the counter is not zero, the sequence cycle time is deducted from the resource counter. If the result becomes zero it means that the process of the present operation is finished. In this case the priority field (PR) of this operation is marked by X, and the priorities of all operations with this operation number are changed to 00.

Automatically the next operation on that item becomes free and gets priority in processing if it is economical to do so. This means that this resource is the "BEST" for this operation or that it's processing time/cost minus a transfer penalty is equivalent to or lower than the "BEST" time of that operation. The operation is allocated to that resource and its resource status file is updated and its counter is set to the new operating time.

As an example: Table 3.5 represents the shop floor status at a certain time. Operation 20 of item #7 was just finished, it was processed on R2, and operation 30 became free. The best resource for this operation is R4 with 7.23 min per item. A check is made if it is economical to process this operation on R2 in order to save transfer time. The time to process on R2 is 8.92. The increase in time is $8.92 - 7.23 = 1.79$. Assume a transfer penalty of 25 min and a quantity of 40 units, then the increase in time is $40*1.79 = 71.6$ and the savings will be only 25 min, which is not economical.

Another case: Operation 20 of item #9 was just finished, it was processed on R3, and thus operation 30 became free. The best resource for this operation is R4 with 11.4 min per item. A check is made if it is economical to process this operation on R3 in order to save transfer time. The time to process on R3 is 11.9 min. The increase in time is $11.9 - 11.4 = 0.4$. Assume a transfer penalty of 25 min and a quantity of 40 units; the increase in time is $40*0.4 = 16$ min and the saving will be 25 min, then it is economical and R3 will process operation 30.

If it is not economical to process the following operation on the previous resource, or if the resource was idle from the previous sequential cycle, then the system scans the matrices of all parts in this particular resource column, and lists all free operations with a "best" mark on them. This list includes all free operations that the specific resource can do best.

If the list contains only one entry, then this entry (operation) is allocated to the resource and its resource status file is updated and its counter is set to the new operating time.

If the list contains more than one entry, then the system allocates the operation with the biggest time gap of performing it on another resource. This value is determined by scanning the operation row in the relevant matrix, and computing the processing time difference between the best resource and the processing time on different resources. Each free operation will be tagged by this difference value. The free operation with the highest tag value will be the one that will be allocated on this sequence cycle on the idle resource.

3 Shop Floor Control

Table 3.6 Status when R5 is idle

Op	PR	R1	R2	R3	R4	R5 IDLE	R6	BEST
	I	T	E	M		#3		
10	X	12.5	9.51	5.15	99	4.02	6.54	5
20	X	5.04	3.93	2.55	99	99	2.82	3
30	X	6.28	4.86	2.98	2.53	2.47	3.44	5
40	*00*	*6.38*	*6.12*	*7.05*	*5.78*	*5.93*	*6.83*	*4*
50	40	8.24	6.33	3.67	2.96	2.62	4.42	5
60	50	5.15	99	4.02	4.86	2.98	2.53	6
	I	T	E	M		#5		
10	*X*	*3.12*	*3.17*	*4.02*	*3.27*	*99*	*99*	*1*
20	*00*	*13.9*	*10.3*	*10.8*	*9.95*	*12.5*	*99*	*4*
30	20	4.86	2.98	2.53	4.86	2.98	2.53	3
40	20	6.04	4.68	2.90	99	99	3.32	3
50	40	5.76	4.47	2.8	99	99	3.18	3
	I	T	E	M		#7		
10	X	3.12	3.17	4.02	3.27	99	99	1
20	X	6.15	4.2	8.05	9.3	99	99	2
30	*00*	*8.34*	*8.92*	*7.58*	*7.23*	*8.76*	*8.12*	*4*
40	30	2.06	2.11	2.96	2.21	99	99	1
	I	T	E	M		#9		
10	X	4.6	3.60	2.39	99	2.05	2.60	5
20	X	5.96	4.59	2.87	99	99	3.28	3
30	*00*	*11.5*	*12.8*	*11.9*	*11.2*	*13.1*	*99*	*4*
40	30	99	99	99	99	1.45	1.72	5
Resource status file								
Res.	Item		Op.	Q		Link	Counter	Hist.
R4	#03		40	60		22	25	66
R1	#7		30	100		23	87	68

Table 3.5 demonstrates this algorithm: R4 is the idle resource and there are four free operations for which this resource is the best one. The system scans these operations across all resources and computes the difference between the minimum time (BEST) and the time on each resource. The maximum difference value is on the column marked by △. In this case the difference between the BEST resource and the resource processing time of item 5 operation 2 is the biggest (13.9 − 9.95 = 3.95). Therefore, this operation will be allocated to the R4 resource. Its resource status file is updated and its counter is set to the new operating time.

If the list is empty a "look ahead" feature is used to determine the "waiting time" for a best operation to become "free". This search is done by scanning the idle resource column for a search for a free operation. When such an operation is encountered, (it is not the best for that resource) the BEST field of this row indicates which resource is the best for that operation. The entry in the field *resource counter* of the *resource status file* indicates the waiting time of that resource.

An example of this procedure is shown by Table 3.6 which shows the status of the 3D matrix at this stage. R5 is idle and searches for a free operation. The free operations are (PR = 00). Scanning the "BEST" column of the table finds that none of the free operations calls for resource R5. The BEST for the free item #3 operation 40 is resource R4. Calling the *resource status file* in resource R4 row indicates that operation 40 is in process and it will take another 25 min to end, which means that waiting time for operation 40 is 25 min.

The system checks if it will be economical to use the idle resource to process the free operation. One method is to compute the difference in time between the BEST and the alternate operation, and compare it to the time that the free resource would otherwise be idle. If the time spent is lower than the time gained, it is economical to do so. The computation is as follows:

Processing the free operation, item #3 operations 40 is by resource R4, and it takes 5.78 min per unit. However resource R4 will become idle only after 25 min. Processing this operation on the idle resource R5 takes 5.93 min per unit. Suppose that the quantity is 100 units, then by working "inefficiently" and increasing the processing time by $(5.93 - 5.78) * 100 = 15$ min gives a savings of $(25 - 15) = 10$ min in throughput time.

Checking the other three open operations indicates that this is the best alternative. Therefore item #3 operation 40 is loaded on R5.

If the finished operation was the last one in processing an item, the data of that item is removed from the 3D matrix, calling the bill-of-material for another item. The new item data (item name and quantity) are recorded and its process plan from the two-dimensional matrix master file is introduced into the 3D matrix.

In case of disruption; the *finish time* on the *history file* will list the time of the interrupt, the *resource counter* of the *resource status file* will be set to 99, which will be set back to zero when the resource is in working condition again. A new job for that operation (item and operation number) is opened with the remaining quantity. This procedure is for a single or multi-resource disruption.

3.3 Summary

The job release stage was done in an office with stable conditions. However, conditions on the shop floor are dynamic. Therefore, the decisions on the shop floor must consider the immediate shop floor status, adding flexibility and dynamics in the shop floor control.

The proposed system is a shop floor control method that does not plan in advance the routine for each released job, therefore bottlenecks cannot be created and disruptions are solved automatically. It is allowed to alter the process when necessary.

The matrix method is a tool that can generate a process, considering the immediate state of the shop floor, and do it within a split second.

Table 3.7 Comparison of scheduling strategies

Optimization criteria	No. periods to process	Unit cost
Maximum production	35	162
Minimum cost	32	76.2
Semi flexible	23	131
Outmost strategy	21	102

To validate the flexibility of the proposed system and to check the execution time, a demonstration program was prepared. The demonstration program can handle several orders and parts. However for simplicity and clarification of the system 2 orders, 12 items, 35 operations, and 15 resources were considered in the example.

Full details are presented in Chap. 4.

Simulation results are shown in Table 3.7.

Chapter 4
Production Planning: Demonstration

Abstract This chapter is an extension of Chap. 3 in which a dynamic system that integrates process planning with production management was presented.

The philosophy of this system is that routing is a variable. A production planner, based on knowledge of customer orders and plant load, decides—at the moment a decision is needed—which order filling route to use. A product tree of each order is used throughout the planning and execution stages.

This chapter demonstrates how the method is used in production planning and scheduling.

1 Introduction

Chapter 3 presented a dynamic system that integrates process planning with production management. The system integrates all stages (i.e. MRP, capacity planning, and shop floor planning and control) with one dynamic logistic all-embracing program. Dynamic and flexible behavior is achieved by employing the roadmap system in process planning. The philosophy of this method is that routing is a variable. The system allows the production planning expert in charge of a given set of orders to decide—at the moment a decision is needed—which route to use in the light of customer orders, their quantities and delivery dates, as well as the current plant load. A product tree of each order is used throughout the planning and execution stages.

The planning system has both finite capacity and access to the product tree of each order. The detailed capacity plan is transformed into a real schedule. The schedule is the basis for dispatching jobs to the shop floor. However, the shop floor is free to produce the released product mix by any other routing and resources in order to solve problems caused by disruption, bottlenecks, resource utilization, etc. with the restriction that the released product mix, as specified for the period, must be completed on time.

The planning steps are as follows:

1. Determination of stock allocation priorities
2. Stock allocation
3. Adjust quantities by economic considerations
4. Capacity planning—resource loading
5. Job release for execution
6. Shop floor control

To validate the feasibility of the proposed system and to check the execution time, a simulation and demonstration computer program was prepared. The program is designed to handle a number of orders and parts. The product bill-of-materials includes all product items, both manufactured and purchased. The purchased items are treated by the purchasing department in the usual manner, and the manufacturing items are treated by our system.

1.1 The Scenario

The demo assumes a company that manufactures metal parts and products. For simplicity and clarification of demonstration two products are ordered:

The orders are:

120 units of product #1 to be delivered on day 40,
40 units of product #11 to be delivered on day 35.

Their bill-of-materials is shown in Fig. 4.1. Each one of the products is of four levels, 10 parts, with several parts common to both of them.

A basic process plan BP is made for each product as shown in Table 4.1. It indicates the type of operation—priority of the operation, length of operation (length of cut), depth of cut, feed rate, cutting speed—and computes required power and processing time. The full BP record includes all the details that are needed to convert the BP process plan to a practical process plan for each resource, and constructing the matrix.

Examining the BP as presented in Table 4.1 reveals that there are several types of operations for an item. Not all operations can be done with the same clamping or the same resource. Whenever an item in an intermittent stage has to be processed on a different resource, it is regarded as a unique item on the bill-of-materials. Therefore the bill-of-materials as shown in Fig. 4.1 is a working bill-of-materials. The difference between a product bill-of-materials and the working bill-of-materials is due to the stages of the process plan. The working bill-of-materials considers each processing stage as an item. The stages are necessary, because otherwise the part cannot be processed.

For example: whenever an item has to be machined from both sides, it must be removed from the resource, turned over and re-clamped, the same as if a different resource is needed for each operation.

1 Introduction

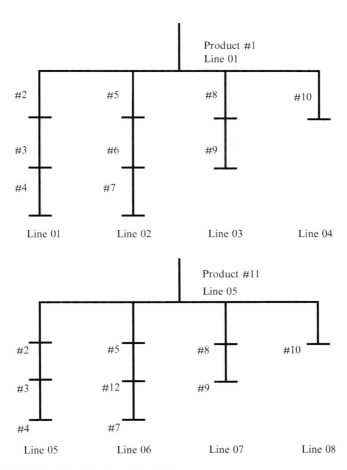

Fig. 4.1 Bill-Of-Materials of product #1 and #2

The first operation on a part is to prepare the raw material. It is a sawing operation, probably done in the store room. Therefore, it appears in the working bill-of-materials as a separate branch and item #7. The part is clamped in the milling resource on one surface, while machining the parallel surface (item #6). Then the item is removed from the resource, and is clamped on the machined surface to assure parallelism, after which the final machining of an accurate side takes place. Only then is the part ready for assembly and is called item #5.

Brief descriptions of the 15 resources which are considered in this demonstration are detailed in Table 4.2. The full resource record includes all the details of the resource needed for transformation of the BP to a resource-operation time matrix.

The BP process plan for all items is transformed into a roadmap matrix as shown in Tables 4.3 and 4.4. The transformation checks the available power, forces, speeds and feeds of the resource to those required by the BP, if the required specifications

Table 4.1 BP process for product #11

	Operation	No.	PR	Tool	L	a	Feed		KW	T Min
Cut	Material	Raw		Item	#7					
010	Sawing	10	00	band						2.8
Mill	External	Side		Item	#6					
010	Side milling	10	00	50	113 × 2	2.5	304	88	1.0	2.02
020	Face milling—R	20	10	80	76	1.59	1,428	118	6.63	0.05
030	Face milling—F	30	20	80	143	0.41	900	147	1.34	0.16
040	Assembly Holes	40	10	10	20 × 4		0.2	25	0.6	0.96
Mill	Internal	Side		Item	#5					
010	Side milling	10	00	50	113 × 2	2.5	304	88	1.0	2.02
020	Face milling—R	20	00	80	76	1.79	1,401	116	7.34	0.05
030	Face milling—F	30	20	80	143	0.21	370	164	0.37	0.39
040	Drill U 46∅	40	10	44	40		0.27	114	10.3	0.2
050	Boring 46∅—Mr	50	40	40	40		804	93	0.22	0.18
060	Boring 46∅—Mf	60	50	40	40		359	106	.07	0.41
070	piston hole	70	60	16	16				2.0	0.34

Table 4.2 The available resources

No	Resource description	Power KW	Speed RPM	Handle time, min	Relative cost
1	Milling Machining center	35	1,500	1.10	4
2	Large CNC Milling	35	1,200	1.15	3
3	Manual milling machine	15	1,500	1.66	1.4
4	Small drill press	2.5	1,200	1.50	1
5	Old milling machine	15	2,400	2.00	1
6	Small CNC milling	10	3,000	1.25	2
7	CNC Lathe	25	3,000	1.15	3
8	Manual Lathe new	15	3,000	1.42	2
9	Manual Lathe old	10	2,400	1.66	1
10	Circular Saw				1
11	Band Saw				1
12	Hack Saw				1
13	Manual assembly				1
14	Machine assembly				1.5
15	Robotics assembly				3

are available only then is the resource handling time added to the BP operation processing time. Examining the BP process as shown in Table 4.1 reveals two important points:

1. The BP process must not necessarily be theoretical. All operations are limited by technology and the required metal removal dimensions, and must actually fit without any changes to the available resources. Actually, the maximum BP power is 10.3 kW, and the maximum available on the resources is 35 kW. Therefore, the only translation needed is to add the handling time. Therefore, the matrix time for operation 010 of item #5 will be:

1 Introduction   77

Table 4.3 Time content of the matrices

OP	PR	R#1	R#2	R#3	R#4	R#5	R#6	R#7	R#8	R#9	#10	#11	#12
		I	T	E	M			#4					
10	00	99	99	99	99	99	99	99	99	99	7.8	6.6	12.2
		II	T	E	M			#3					
10	00	3.12	3.17	3.68	99	4.02	3.27	99	99	99	99	99	99
20	10	1.15	1.2	1.71	99	2.05	1.3	99	99	99	99	99	99
30	20	1.26	1.31	1.82	99	99	1.41	99	99	99	99	99	99
40	10	1.57	1.62	2.13	2.53	2.47	1.72	99	99	99	99	99	99
50	10	2.06	2.11	2.62	2.62	2.96	2.21	99	99	99	99	99	99
		I	T	E	M			#2					
10	00	3.12	3.17	3.68	99	4.02	3.27	99	99	99	99	99	99
20	00	1.15	1.2	1.71	99	2.05	1.3	99	99	99	99	99	99
30	20	1.49	1.53	2.05	99	99	1.64	99	99	99	99	99	99
40	10	1.30	1.35	1.86	1.86	2.2	1.45	99	99	99	99	99	99
50	40	1.28	1.33	1.84	99	2.18	1.43	99	99	99	99	99	99
60	50	1.51	1.56	2.07	99	99	1.66	99	99	99	99	99	99
		I	T	E	M			#7					
10	00	99	99	99	99	99	99	99	99	99	7.8	6.6	12.2
		I	T	E	M			#6	&	#12			
10	00	3.12	3.17	3.68	99	4.02	3.27	99	99	99	99	99	99
20	10	1.15	1.2	1.71	99	2.05	1.3	99	99	99	99	99	99
30	20	1.26	1.31	1.82	99	99	1.41	99	99	99	99	99	99
40	10	2.06	2.11	2.62	2.62	2.96	2.21	99	99	99	99	99	99
		I	T	E	M			#5					
10	00	3.12	3.17	3.68	99	4.02	3.27	99	99	99	99	99	99
20	00	1.15	1.2	1.71	99	2.05	1.3	99	99	99	99	99	99
30	20	1.49	1.53	2.05	99	99	1.64	99	99	99	99	99	99
40	10	1.30	1.35	1.86	1.86	2.2	1.45	99	99	99	99	99	99
50	40	1.28	1.33	1.84	99	2.18	1.43	99	99	99	99	99	99
60	50	1.51	1.56	2.07	99	99	1.66	99	99	99	99	99	99
70	60	1.44	1.49	2.00	99	99	1.59	99	99	99	99	99	99
		I	T	E	M			#9					
01	00	99	99	99	99	99	99	2.03	2.33	2.57	99	99	99
20	10	99	99	99	99	99	99	1.91	1.85	99	99	99	99
		I	T	E	M			#8					
10	00	99	99	99	99	99	99	1.45	1.72	1.96	99	99	99
20	10	99	99	99	99	99	99	1.45	1.72	99	99	99	99
		I	T	E	M			#10					
10	00	99	99	99	99	99	99	2.86	3.12	99	99	99	99

On resource $1 - 2.02 + 1.1 = 3.12$

On resource $2 - 2.02 + 1.15 = 3.17$

On resource $3 - 2.02 + 1.66 = 3.68$

Table 4.4 Cost content of the matrices

OP	PR	R#1	R#2	R#3	R#4	R#5	R#6	R#7	R#8	R#9	#10	#11	#12
		I	T	E	M			#4					
10	00	99	99	99	99	99	99	99	99	99	7.8	6.6	12.2
		II	T	E	M			#3					
10	00	12.5	9.51	5.15	99	4.02	6.54	99	99	99	99	99	99
20	10	4.60	3.60	2.39	99	2.05	2.60	99	99	99	99	99	99
30	20	5.04	3.93	2.55	99	99	2.82	99	99	99	99	99	99
40	10	6.28	4.86	2.98	2.53	2.47	3.44	99	99	99	99	99	99
50	10	8.24	6.33	3.67	2.62	2.96	4.42	99	99	99	99	99	99
		I	T	E	M			#2					
10	00	12.5	9.51	5.15	99	4.02	6.54	99	99	99	99	99	99
20	00	4.4	3.60	2.39	99	2.05	2.60	99	99	99	99	99	99
30	20	6.0	4.59	2.87	99	99	3.28	99	99	99	99	99	99
40	10	5.2	4.05	2.60	1.86	2.2	2.90	99	99	99	99	99	99
50	40	5.12	3.99	2.58	99	2.18	2.86	99	99	99	99	99	99
60	50	6.04	4.68	2.90	99	99	3.32	99	99	99	99	99	99
		I	T	E	M			#7					
10	00	99	99	99	99	99	99	99	99	99	7.8	6.6	12.2
		I	T	E	M			#6	&	#12			
10	00	12.5	9.51	5.15	99	4.02	6.54	99	99	99	99	99	99
20	10	4.60	3.60	2.39	99	2.05	2.60	99	99	99	99	99	99
30	20	5.04	3.93	2.55	99	99	2.82	99	99	99	99	99	99
40	10	8.24	6.33	3.67	2.62	2.96	4.42	99	99	99	99	99	99
		I	T	E	M			#5					
10	00	12.5	9.51	5.15	99	4.02	6.54	99	99	99	99	99	99
20	00	4.6	3.60	2.39	99	2.05	2.60	99	99	99	99	99	99
30	20	5.96	4.59	2.87	99	99	3.28	99	99	99	99	99	99
40	10	5.20	4.05	2.60	1.86	2.20	2.90	99	99	99	99	99	99
50	40	5.12	3.99	2.58	99	2.18	2.86	99	99	99	99	99	99
60	50	6.04	4.68	2.90	99	99	3.32	99	99	99	99	99	99
70	60	5.76	4.47	2.8	99	99	3.18	99	99	99	99	99	99
		I	T	E	M			#9					
01	00	99	99	99	99	99	99	6.09	4.66	2.57	99	99	99
20	10	99	99	99	99	99	99	5.73	3.70	99	99	99	99
		I	T	E	M			#8					
10	00	99	99	99	99	99	99	4.35	3.44	1.96	99	99	99
20	10	99	99	99	99	99	99	4.35	3.44	99	99	99	99
		I	T	E	M			#10					
10	00	99	99	99	99	99	99	8.58	6.24	99	99	99	99

The matrix time for operation 07 of items #5 will be:

On resource 1 − 0.34 + 1.1 = 1.44

On resource 3 − 0.34 + 1.66 = 2.00

On resource 6 − 0.34 + 1.25 = 1.59

The matrix for operation 4 of item #5 requires 10.3 kW, which is available on all resources except for resource 4; it will be computed in a similar way, i.e. add 0.2 to the handling time of each resource.

2. Operation 04 of item #5 requires 10.3 kW, however resource 4 has only 2.5 kW. This operation (see Table 4.1) is drilling a 44 mm diameter hole and is done with an insert drill of 44 mm diameter. To reduce the required power, this hole will be made in two steps, first drill to 20 mm diameter, and then increase the hole to 44 mm diameter. The 20 mm diameter with an insert drill will take 0.18 and require 2.47 kW. The second operation will take 0.17.

The total hole-making item #5 operation 4 is then:

0.19 + 0.17 = 0.36 plus the handling time of 1.50 total 1.86.

This value is inserted in the matrix.

The content of the roadmap gives the time to perform each operation on each resource. In case that the resource cannot be used to process an operation the time is designated by high value, (99 in Tables 4.3 and 4.4). The same mathematical solution of the matrix is used for the cost matrix (shown in Table 4.4) as well as the time matrix (shown in Table 4.3). However, regardless of the criterion of optimization used in solving the matrix, the selected process is always given, in the case of capacity planning, by the time to process an operation.

Note: Due to space restrictions the matrices Tables 4.3 and 4.4 omit the data for resources 13, 14, 15. They are used, in this example, only for assembly of products 1 and 11.

Their data is:

R13	Manual assembly	time is 5.0	cost 5.0
R14	Machine assembly	time is 4.0	cost 6.0
R15	Robotic assembly	time is 3.0	cost 9.0

2 The Planning Steps

The planning steps are:

1. Determination of stock allocation priorities
2. Stock allocation

3. Adjust quantities by economic considerations
4. Capacity planning—resource loading
5. Job release for execution
6. Shop floor control

2.1 Determination of Stock Allocation Priorities

The objective of this stage is to set priorities for stock allocation. The strategy is to allocate the stock to the critical order, where critical is defined as the order that its lowest level item has to start at the earliest time. The earliest time might be in the past, or in the future, or exactly at the current day of the planning date. To meet this strategy the first step is to build the product tree on a time-based scale, instead of on levels.

The time-based product tree is constructed from the order delivery date backwards. The name of the order is retrieved from the level-based product tree (level 0) and calls the matrix to generate a process plan. The processing time of such a process is given for a single item. Compute the total processing time, by multiplying the order quantity by the process time of a single product.

Next each item of level 1 for the same order is treated individually using the starting time of level 0 as the delivery date for each item of level 1. The starting date of the level 1 item is the due date for level 2 items and so forth.

This process is repeated for all levels of the order and for all orders in the file.

2.1.1 Time-Based Product Tree

The computation starts with the assembly of order number one, i.e. item #1. The matrix is called to generate a process for the assembly of 120 items of order 1; assembly #1. The output of the matrix module gives the total time, and the detailed operations, including the time on each individual resource. For demonstration purposes assume a constant setup and other expenses of 30 min and unit cost.

Therefore the penalty is $30/120 = 0.25$.

The matrix recommends using resource 11.
The time is 5.0 min $+ 30/120 = 5.25$ min.
Total time to process 120 items is $(120 \times 5.25) = 630$ min.
For a day of 480 min it is 1.31 days.
The delivery date is day 40.
Therefore the assembly must start on day $(40 - 1.31) = 38.69$.

The next item on the bill-of-materials (see Fig. 4.1) is item #2. The quantity per unit assembly is 1. Therefore the quantity of item #2 is 120 units.

The matrix is called to generate a process for processing 120 pieces of item #2. The recommended process is:

Resource 5 operations 10, 20	4.02 + 2.05 =	6.07 + 0.25 =	6.32 min
Resource 3, operation 30		2.05 + 0.25 =	2.30 min
Resource 4, operation 40		1.86 + 0.25 =	2.11 min
Resource 5, operation 50		2.18 + 0.25 =	2.43 min
Resource 3, operation 60		2.07 + 0.25 =	2.32 min
Total			15.48 min

For 120 items × 15.48 = 1857.6 min/480 min/day = 3.87 days.

Item finish date is 38.69 therefore, it must start on day 38.69 − 3.87 = 34.82.

Next on the bill-of-materials is item #3. The quantity per unit assembly is 1. Therefore the quantity of item #3 is 120 items. The matrix is called to generate a process for processing 120 pieces of item #3. The output of the matrix module is to use:

Resource 5 operations 10, 20, 40	4.02 + 2.05 + 2.47 =	8.54 + 0.25 =	8.79 min
Resource 3 operation 30		1.82 + 0.25 =	2.07 min
Resource 4 operation 50		2.62 + 0.25 =	2.87 min
Total			13.73 min

For 120 items × 13.73 = 1647.6 min/480 min/day = 3.432 days.

Item finish date is 34.82 therefore, it must start on day 34.82 − 3.43 = 31.38.

Next on the bill-of-materials is item #4. The quantity per unit assembly is 1. Therefore the quantity of item #4 is 120 items. The matrix is called to generate a process for processing 120 pieces of item #4. The output of the matrix module is to use:

Resource 11 operation 10 6.6 + 0.25 = 6.85 min

For 120 items 120 × 6.85 = 822 min which are 822/ 480 = 1.71 days.

Item #4 must be finished on day 31.38, therefore it must start on day 31.38 − 1.71 = 29.67.

This branch is finished.

Similar computations as detailed above are repeated for the next branch of items #5; #6; #7 and followed by the branch of items #8; #9. Both must be ready for assembly on day 38.69. The computations of the starting date continue to the second order as well.

The working product tree of the orders is shown in Fig. 4.2.

2.2 Stock Allocation

The objective of this step is to allocate available stock and to adjust the quantities of each item on the ordered products tree. The term "available stock" includes items in the store room, on the shop floor, in a purchasing process, in inspection and etc. The strategy is to allocate the available stock to the critical order, where critical is

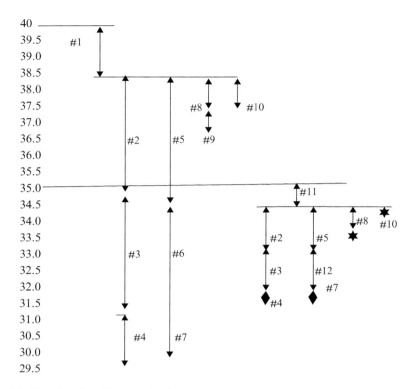

Fig. 4.2 Time-based product tree of orders

defined as the order that its lowest level item has to start at the earliest time. The allocation is carried out with the working product tree, as constructed and shown in Fig. 4.2 and gives priority to the critical order. The program scans all branches of the working bill-of-material in a search for a low level item of which the starting day is the earliest one.

In this example, as indicated in Fig. 4.2 it is item #4 of order 1 that has the starting date of 29.67 (the same results are displayed in a table format for this first step, for following steps see Tables 4.5 and 4.6). Therefore, order 1 is the critical one and stock allocation will start with this order (see Fig. 4.1). It starts with the high level items and proceeds towards the low level items. Hence, initially product #1 will search for available stock. In this example the available stock shows:

Item	#1	#4	#5	#6	#8	#9	#10	#11	#12
Quantity	40	10	18	9	11	20	28	20	15

2 The Planning Steps   83

Table 4.5 Product tree in table format: stock allocation

			First step		Second step		Third step	
Order	Due date	Itm	Quantity	Start date	Quantity	Start date	Quantity	Start date
1	40		120					
		#1	120	38.69	80	39.10	80	39.10
		#2	120	34.82	80	36.55	80	36.55
		#3	120	31.38	80	34.20	80	34.20
		#4	120	**29.67**	80	33.04	80	**33.04**
		#5	120	34.32	80	36.22	80	36.22
		#6	120	31.50	80	34.28	80	34.28
		#7	120	29.79	80	33.12	80	33.12
		#8	120	37.64	80	38.37	80	38.37
		#9	120	36.41	80	37.50	80	37.50
		#10	120	37.85	80	38.52	80	38.52
2	35		40					
		#11	40	34.52	40	34.52	20	34.73
		#2	40	33.24	40	33.24	20	34.12
		#3	40	31.97	40	31.97	20	33.56
		#4	40	31.36	40	**31.36**	20	33.22
		#5	40	33.07	40	33.07	20	34.03
		#12	40	32.01	40	32.01	20	33.56
		#7	40	31.40	40	31.40	20	33.22
		#8	40	34.09	40	34.09	20	34.52
		#9	40	33.60	40	33.60	20	34.21
		#10	40	34.20	40	34.20	20	34.54

Therefore, the available stock of 40 items will be allocated to item #1. The order is for 120 units, thus $120 - 40 = 80$ units are needed to meet the original order. As it is level 1, all the following items of the working bill-of materials should be adjusted accordingly.

After each allocation the working bill-of-material changes, and thus the critical order is changed.

The matrix will be called to generate a process plan for the new quantity, in the same manner as before. These results are shown in Table 4.5, step 2.

In this example, as indicated in Table 4.5, second step, the critical path is changed, the earliest starting date is now for item #4 of order 2 (day 31.36). Following the bill-of-materials tree, from this item onwards, indicates that this item is part of product #11, order number 2.

Therefore, order 2 is the critical one and stock allocation will start with this order. Hence, initially product #11 will search for available stock. There are 20 products in stock and they are allocated to order 2. Therefore, the quantity for item #11 will be reduced to 20 products and the quantities of all items in the product tree will be adjusted accordingly.

Table 4.6 Product tree in table format: stock allocation

Order	Due date	Itm	Fourth step Quantity	Start date	Fifth step Quantity	Start date	Final step Quantity	Start date
1	40		120					
		#1	80	39.10	80	39.10	80	39.10
		#2	80	36.55	80	36.55	80	36.55
		#3	80	34.20	80	34.20	80	34.20
		#4	80	33.17	70	33.17	70	33.17
		#5	80	36.22	62	36.93	62	36.93
		#6	80	34.28	53	35.58	53	35.58
		#7	80	33.12	53	34.79	53	34.79
		#8	80	38.37	69	38.45	69	38.45
		#9	80	37.50	49	37.87	49	37.87
		#10	80	38.52	52	38.70	52	38.70
2	35		40					
		#11	40	34.73	20	34.73	20	34.73
		#2	20	34.12	20	34.12	20	34.12
		#3	20	33.56	20	33.56	20	33.56
		#4	20	33.22	20	33.22	20	33.22
		#5	20	34.03	20	34.03	20	34.03
		#12	20	33.56	20	33.56	5	33.87
		#7	20	33.22	20	33.22	5	33.74
		#8	20	34.52	20	34.52	20	34.52
		#9	20	34.21	20	34.21	20	34.21
		#10	20	34.54	20	34.54	20	34.54

The matrix will be called to generate a process plan for the new quantity in the same manner as before. This state is shown in Table 4.5 step 3.

The program again scans all paths of the bill-of-material trees in a search for a low level item of which the starting day is the earliest one. In this example, as indicated in Table 4.5, third step, the critical path is changed, the earliest starting date is now for item #4 of order 1 (day 33.04). Following the bill-of-materials tree, from this item onwards, indicates that this item is part of product 1, order number 1.

Therefore, order 1 is the critical one and stock allocation will start with this order. Hence, initially product 1 will search for available stock. As item #1 was already allocated, the next item of product 1 is item #2. As there is no item #2 in stock, the bill-of-materials proceeds to item #3. As there is no stock for item #3, the bill-of-materials proceeds to item #4. There are 10 units of item #4 and they are allocated to item #4 of order 1. The quantity of this item is reduced by 10–70 required units.

The matrix is called to generate a process for 70 items #4 and the starting day is adjusted accordingly, as shown in Table 4.6, fourth step.

The program again scans all paths of the bill-of-materials trees in a search for a low level item of which the starting day is the earliest one. In this example, as indicated in Table 4.6 fourth step, the critical item is item #7 of order 1 (day

33.12). Following the bill-of-materials tree, from this item onwards, indicates that this item is part of product 1, order number 1 with the path of items #7; #6; #5 & #1. Therefore, order 1 is the critical one and stock allocation will start with this order. Item #1 is marked as treated therefore the bill-of-materials points to item #5. As there are 18 units of item #5 in stock they are allocated to item 5 of order 1. The quantity of item 5 is reduced to 62 units.

The matrix is called and generates a process to produce 62 units of item #5.

Item #4 of order 1 is the earliest starting day, order 1 remains the critical path, and the allocation proceeds accordingly, as can be seen in Table 4.6 fifth step. Then stock allocation proceeds to order 2. The final state after stock allocation is shown in Table 4.6 final step.

The critical order as shown in the final step will be used to set the capacity planning of the next stage.

2.3 Capacity Planning: Resource Loading

The objective of this stage is to set priorities for jobs to be released to the shop floor for processing. Other objectives are intended for management miscellaneous objectives and are not discussed in this section. This section presents the effect of process planning criterion of optimization on the capacity planning outcome. Several options are presented. Please remember that the real resource loading is carried out on the shop floor, and there the loading strategy is different.

The working product structure lists all items that must be processed on the shop floor. It details the required quantify of each item, and its relationship to other items. The list is detailed in Table 4.6, the final step. The starting day on the list indicates the priorities of resource loading. The item with the earliest starting date points to the order which is the critical order; therefore it should be loaded first. The loading is forward, from the low level item toward the first sub-assembly. Then all other items that are needed for the assembly are loaded, and so on.

In the example, and the final working product structure, item #4 of order 1 is the item with the earliest starting date (33.17). Therefore it will be the first item to be loaded.

The sequence of items to be loaded is as follows:

Order 1—items 4; 3; 2; 7; 6; 5; 9; 8; 10; assemble item 1.

Next order 2 is treated, the earliest start day is on item 4 (33.22) therefore the sequence will be: items:

Order 2—items 4; 3; 2; 7; 12; 5; 9; 8; 10; assemble item 11.

In the experimental computer program an array of resources was used and each period was of 120 min.

The entry in each array location indicates if the resource is idle or occupied. A blank value indicates an idle period, otherwise the item and order number that occupy the resource at that period is listed. In this example the first digit indicates the item number as indicated in Fig. 4.1 and the last two digits indicate the line number.

Initially, the items that are in operation on the shop floor are listed in the array. This is because once processing of an item starts it should be completed. The capacity planning—resource loading in this example-assumes that there are no in-processing items, and the resource-period array is all blanks. Several loading strategies are evaluated.

2.3.1 Capacity Planning: Minimum Cost Process Plan

The critical order is order 1 and the critical item is *item #4*. The quantity required is 70 units. The matrix is called to generate a process for 70 units of item #4 (therefore the penalty is $30/70 = 0.428$).

The recommended process plan is to use:

$$\text{Resource 11 for } 6.6 + 0.428 = 7.03 \text{ min. times 70 PCs.}$$
$$= 491.96 \text{ min. divided by a period of 120 min.} = 4 \text{ periods.}$$

As this operation can be made at any period, the system scans resource 11 and finds that period 1 is available (empty) for four periods. Therefore, the first four periods are allocated to perform item 4 of order 1, marked 401 on Table 4.7.

The next *item #3* of order 1 is treated. The quantity is 80 units (therefore the penalty is $30/80 = 0.375$). The matrix is called and a minimum cost process plan is generated, it calls for:

$$\text{Resource 5 for operations 1; 2; 4 total of 8.54 minutes} + 0.375$$
$$\text{Resource 3 for operation 3} \quad 1.82 + 0.375 \text{ minutes}$$
$$\text{Resource 4 for operation 5} \quad 2.62 + 0.375 \text{ minutes}$$

The loading starts with resource 5: $80 \times (8.54 + .375)/120 = 5.94 = 6$ periods.

The loading may start only after item #4 is finished, i.e. period 5. As resource 5 is available at those periods it is loaded at period 5–10 and marked as 301 on Table 4.7.

Next resource 3 for:
$80 \times (1.82 + .375)/120 = 1.46 = 1$ period (the loadings are in rounded integers).

The loading may start only after resource 5 finishes i.e. period 11. Therefore period 11 on resource 3 is marked by 301.

Next resource 4 for: $80 \times (2.62 + .375)/120 = 1.99 = 2$ periods.

The loading may start only after a resource's finish, i.e. period 12.

Table 4.7 Load profile for minimum cost process plan

	R1	R2	R3	R4	R5	R6	R7	R8	R9	R10	R11	R12
1								1,008	903		401	
2								903	907		401	
3								907	803		401	
4								1,004			401	
5					301			1,004			702	
6					301			807			702	
7					301						702	
8					301						405	
9			305		301						706	
10			305		301							
11			301		602							
12			205	301	602							
13			205	301	602							
14			602		201							
15			1,206	602	201							
16			506		201							
17			506		201							
18			506		201							
19					201							
20					201							
21			201		502							
22			201		502							
23			201		502							
24			502									
25			502									
26			502									
27			502									
28			502									
29												

As resource 4 is available at period 12 for a duration of two periods, it is loaded and marked as 301.

Next *item #2* of order 1 is treated. The quantity is 80 units. The matrix is called and a minimum cost process plan is generated, it calls for:

Resource 5 for operations 1; 2; 4; 5 total of 10.45 minutes + 0.375
Resource 3 for operation 3; 6 total 4.12 + 0.375 minutes

The loading starts with resource 5: $80 \times (10.45 + .375)/120 = 7.21 = 7$ periods. The loading may start only after item 3 is finished, i.e. period 14. As resource 5 is available at those periods it is loaded at period 14–20 and marked as 201.

Next resource 3 for: $80 \times (4.12 + .375)/120 = 3$ periods.

The loading may start only after resource 5 finishes i.e. period 20. Therefore periods 21; 22; 23 on resource 3 are marked by 201.

Next item #7 of order 1 is treated. The quantity is 53 items (therefore the penalty is 30/53 = 0.566). The matrix is called and a minimum cost process plan is generated, it calls for:

Resource 11 for: 6.6 min + 0.566. i.e. it requires $53 \times (6.6 + 0.566)/120 = 3.16 = 3$ periods. This operation may start at period 1. However, resource 11 is occupied at periods 1–4 (these periods are marked by 401) and the earliest period is period 5. Therefore, item 7 is loaded at periods 5, 6, &7 on resource 11 and marked 702.

Next *item #6* of order 1 is treated. The quantity is 53 items. The matrix is called and a minimum cost process plan is generated, it calls for:

Resource 5 for operations 1; 2 total of 6.0 7 minutes + 0.566
Resource 3 for operation 3; 1.82 + 0.566 minutes
Resource 4 for operation 4; 2.62 + 0.566 minutes

The loading starts with resource 5:

$$53 \times (6.07 + .566)/120 = 2.93 = 3 \text{ periods}$$

The loading may start only after item #7 is finished, i.e. period 8. As resource 5 is not available at this period its start is delayed till machine 5 is available, i.e. period 11. The loading of resource 5 with item #6 is therefore on periods 11; 12 & 13, and is marked as 602.

Next resource 3, $53 \times (1.82 + .566)/120 = 1$ period.

The loading may start only after resource 5 finishes i.e. period 14. Therefore period 14 on resource 3 is marked by 602.

Next resource 4, $80 \times (2.62 + .566)/120 = 1.2 = 1$ period.

The loading may start only after a resource's finish, i.e. period 15. As resource 4 is available at period 15 for a duration of one period, it is loaded and marked as 602.

Next *item #5* of order 1 is treated. The quantity is 62 items (therefore the penalty is 30/62 = 0.484). The matrix is called and a minimum cost process plan is generated, it calls for:

Resource 5 for operations 1; 2 total of 6.07 minutes + 0.484
Resource 3 for operation 3; 4; 5; 6 total of 9.82 + 0.484 minutes
Resource 5 $62 \times (6.07 + .484)/120 = 3.38 = 3$ periods.

The loading can start only after item 6 is finished (i.e. period 15). As resource 5 is not available at period 15, the loading of item 5 has to wait till period 21 (i.e. for five periods) as resource 5 is occupied by item 602. The loading will start at period 21 for three periods, 21; 22; 23.

$$\text{Next resource 3} \quad 62 \times (9.82 + .484)/120 = 5.32 = 5 \text{ periods}$$

The loading can start only after resource 5 finishes (i.e. period 23). Therefore period 24; 25; 26; 27; 28 on resource 3 are loaded with item 5 and marked by 502.

Next *items #9; #8* and *#10* are treated and loaded on resource 9 & 8.

Next the assembly of order 1, *item #1* can be made. It calls for resource 13 for $5 + 0.375$ min, $80 \times 5.375/120 = 3.58 = 4$ periods. The resource is available at any time; however, the assembly may start only when all items of the assembly are available. The last item available is item #5 on period 28. Therefore, the assembly will be made on periods 29; 30; 31 & 32 and mark resource 13 at these periods by 101.

Next order 2 is treated in the same manner. The loadings of all items are shown in Table 4.7.

Please note that operation 405 may start at period 1 however, due to resource loading it starts at period 8. Item #3 of order 2 may start after item #4 of order 2. It requires resource 3 for two periods. (Note that due to the low quantity, and the high penalty of changing resources, the matrix recommends to use only resource 3, as compared to its recommendations to use resource 5 & 3 for order 1). The processing of item #3 of order 2 may start after item #4 i.e. at period 9. Resource 3 is idle till period 11, when it is occupied by 301. Therefore item #3 of order 2 load resource on periods 9 & 10 are marked by 305. Although order 2 is loaded after order 1 the resource loads it before, due to resource idleness. Similarly item #2 of order 2 requires two periods on resource 3. It may start at period 11, however the resource is occupied by item 301, and it becomes idle at periods 12 and 13 which are loaded and marked by 205.

The assembly of order 2 is on resource 13 at period 19.

2.3.2 Capacity Planning: Maximum Production Process Planning

In this case the matrix is programmed to generate a process plan using the maximum production criterion of optimization. The loading method is the same as before. Examining the general time-based matrix, Table 4.3 shows that the shortest processing time of all milling operations are when resource 1 is used. It is clear that the best resource for each item is selected, i.e. resource 1. The loading profile is shown in Table 4.8.

It is strange that when working with maximum production criteria of optimization the lead time to manufacture the product mix, in this example, is (34 periods)

Table 4.8 Load profile for a maximum production process plan

	R1	R2	R3	R4	R5	R6	R7	R8	R9	R10	R11	R12
1							903				401	
2							903				401	
3							803				401	
4							803				401	
5	301						1,004				702	
6	301						907				702	
7	301						807				702	
8	301						1,008				405	
9	301										706	
10	301											
11	201											
12	201											
13	201											
14	201											
15	201											
16	201											
17	201											
18	602											
19	602											
20	602											
21	602											
22	502											
23	502											
24	502											
25	502											
26	502											
27	502											
28	305											
29	305											
30	205											
31	205											
32	1,206											
33	506											
34	506											
35												

longer than the lead time in the case of using minimum cost criteria of optimization (28 periods). This phenomenon is due to the fact that in the maximum production criteria of optimization, all items are scheduled to be processed on the best resource, i.e. resource 1, as a result a long queue of work piling up.

This supports the thesis that a flexible process plan should be used and the selected process to the load profile adjusted.

Table 4.9 Load profile for cost/production process plan

	R1	R2	R3	R4	R5	R6	R7	R8	R9	R10	R11	R12
1							903	1,004			401	
2							903	1,004			401	
3							803	1,008			401	
4							803				401	
5	301						907				702	
6	301						807				702	
7	301										702	
8	301				602						405	
9	301				602						706	
10	301		1,206		602							
11	201		602									
12	201		506	602								
13	201		506		502							
14	201		506		502							
15	201				502							
16	201		502									
17	201		502									
18	305		502									
19	305		502									
20	205		502									
21	205											
22												
23												
24												
25												

2.3.3 Capacity Planning: Flexible Process Plan

A simple case of process interchange occurs when the criteria of optimization interchange is established: once minimum cost is used, then for the next branch of the product structure the maximum production criteria of optimization is used to generate a process plan by the matrix.

The lead time to produce the required product mix in this case is 21 periods. The load profile in this case is shown in Table 4.9.

2.3.4 Capacity Planning: Variable Process Planning

A more sophisticated capacity algorithm could select a different process plan whenever an item has to wait for a resource. Each plant may use any economic algorithm it chooses. Its choice may be dependent on the number of periods that an item must wait for an idle resource, or the cost differences or a combination of the two.

Table 4.10 Load profile for the case of a variable process plan

	R1	R2	R3	R4	R5	R6	R7	R8	R9	R10	R11	R12
1								1,008	903	702	401	
2								903	907	702	401	
3								907	803	702	401	
4								1,004		702	401	
5	301				602			1,004		706	405	
6	301		305		602	1,206		807				
7	301		305		602	506						
8	301		602			506						
9	301		205	602								
10	301		205		502							
11		201			502							
12		201			502							
13		201	502									
14		201	502									
15		201	502									
16		201	502									
17		201	502									
18												
19												
20												

In this example the objective is to produce the product mix of the two orders in the shortest lead time possible while keeping the processing cost to a minimum (as a second objective). The rule used by the algorithm is to change a process whenever an item has to wait, even for one period. The least cost resource should be selected if it does not affect the lead time. The resulted load profile is shown in Table 4.10. The lead time in this case is 17 periods.

For example: *item #4* of order 1 is on the critical path. Therefore it is loaded on resource 11 for four periods.

To meet the least lead time the system uses the maximum production criteria of optimization to generate the process plan for *item #3*. It calls for resource 1 for six periods, and it may start only after item #4 is done, i.e. period 5. As resource 1 is idle at that period, it was loaded with item #3 and marked 301.

A check is made to determine if resource 2 or 6 or 3 can process this item with the same number of periods. The check reveals that: to produce item #3 on resource 2 will last seven periods, on resource 6 it will last seven periods and on resource 3 it will last eight periods. Therefore, the loading of item 3 is done on resource 1.

Next *item #2* is considered, again with the maximum production criteria of optimization. The matrix recommends using resource 1 for seven periods. It may start only after item #3 is done, i.e. period 11. Resource 1 is idle at that period. However resource 2 can process this item also in seven periods (note that the number of periods is rounded to the closer integer to a period). The hourly rate of resource 2 is 75% of that of resource 1 and as there is no lead time saving by using resource 1, it is decided to use resource 2 for item #2 and the periods 11–17 are marked by 201.

2 The Planning Steps

Next ***item #7*** of order 1 is considered for loading. The matrix recommends using resource 11 for three periods and may start at period 1. It scans resource 11 and finds that the resource will be idle only on period 5. In the present loading strategy, item #7 turned to the matrix, (which blocked resource 11 from being selected) and generated another process. This time the matrix recommended a process to use resource 10. The processing time increased from 6.6 to 7.8, but the processing may start without delay, on period 1. It will take four periods on resource 10, instead of three periods on resource 1. By this change item #7 can be finished three periods ahead.

The next item to be loaded is ***item #6***. The matrix recommends using resource 1 for four periods. It can start only after item #7 is done, i.e. period 5. However resource 1 is occupied till period 10. Therefore, instead of waiting for six periods (5–10) the matrix blocks resource 1 and generates another process which recommends using resource 2, which calls for four periods. Resource 2 is available and periods 5–8 are marked as 602.

The next item to be loaded is ***item #5***. The matrix recommends using resource 1 for five periods. It can start only after item #6 is done, i.e. period 9. However resource 1 is occupied till period 10. Therefore, instead of waiting for two periods (9 & 10) the matrix blocks resource 1 and generates another process which recommends using resource 2, which calls for five periods starting at period 9. However Resource 2 is available for only two periods, and is occupied in periods 11–17. Therefore instead of waiting for nine periods (9–17, it is not allowed to split the process) the matrix blocks resource 2 as well and generates another process that recommends using resource 6, which calls for six periods starting after item #6 is done, i.e. period 9. Resource 6 is idle at those periods and item #5 is loaded and periods 9–15 are marked as 502.

The assembly of product 1 may start only after all components are available. As item #2 is the critical one and is done at period 17, it means that the assembly of product 1 may start only on period 18, and calls for three periods. Item 5 is ready for assembly at period 15, which means that it waits for three periods before it can be used for the assembly. Therefore, the system checks if the cost of processing may be reduced without causing any increase in the lead time. Item #7 is already moved to the first period. Therefore item #6 is being evaluated. The matrix is called to generate a process plan with minimum cost criteria of optimization for ***item #6***. The recommended process calls for using:

Resource 5 operations 1; 2 total of 6.07 minutes + 0.566 = 3 periods
Resource 3 operation 31.82 + 0.566 = 1 period
Resource 4 operation 42.62 + 0.566 = 1 period

In total it takes five periods, which means an increase of one period and reduces the cost from 12 units to 5.3 units (resource 2 = 3 units × 4 periods; resource 5 = 1 unit × 3 periods + 1.4 unit × 1 period + 1 unit × 1 period). Therefore the new recommendation is to ***move item #6*** from resource 2 to resources 5; 3; 4 marking period 5; 6 & 7 of resource 5 as 602; period 8 of resource 3 as 602; and period 9 of resource 4 as 602.

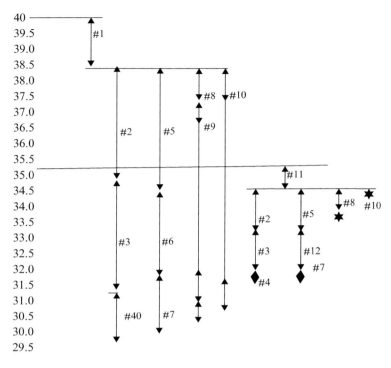

Fig. 4.3 Time-based product tree of orders

A similar treatment to *item #5* indicates that it can use a minimum cost optimization process which calls, as computed before, for three periods of resource 5 and five periods of resource 3. This reduces the cost from using resource 6 ($7 \times 2 = 14$ units; to $3 \times 1 + 5 \times 1.4 = 10$ units). It is economical without effecting the lead time of the assembly, and the change is made, i.e. the load is shifted from resource 6, periods 9–15, and periods 10–12 of resource 5 are marked 502 and periods 13–17 of resource 3 by 502.

The system treats all other items with a similar logic and the results are shown in Table 4.10.

Figure 4.3 shows that there is slack that might be employed to solve scheduling difficulties. Slack is the interval of time between early starting date and early finish date.

2.4 Job Release for Execution

The objective of this stage is to compile a list of jobs that should be processed in the near period. Such a list may be compiled from the capacity planning module as displayed by any matrices in Tables 4.7–4.10 or from any other source.

The user defines the number of periods desired to release for production. The items on the matrices that fall within these periods are recorded on the list.

The list contains the data of:

- Job number and code
- Number of operations
- Quantity
- Preferred resource for the job

Naturally all system files such as Working bill-of-materials (Table 4.2) and time and cost matrices (Tables 4.3 and 4.4) are available at this stage.

The first step is to examine all released jobs with the attempt to **combine batches** of the same job in the released periods if it is economical.

In case that processing a job is complete, it is withdrawn from the list and the next job is added.

2.5 Shop Floor Control

2.5.1 The Strategy

Shop floor control is based on the concept that whenever a resource is free, it searches for a free operation to process. A free operation is identified by scanning the column "PR" of the 3D matrices process (see Tables 4.3 and 4.4). Any operation with $PR = 0$ is a free operation. A free resource is identified by the resource counter as equal to zero (0).

The scheduling module is based on a *sequence cycle loop* that examines all resources listed in the *resource status file*, loads the free resources and updates the *resource counter*.

The *resource* status file keeps the status of the resource throughout the scheduling period. The data stored is:

Resource number
The loaded item
Quantity
A link to the bill-of-materials
Resource counter
Sequence number of entry in the history file.

The resource counter is a counter that indicates the remaining time for processing an item. When loaded it is set by multiplying the quantity by the processing time, as indicated by the 3D matrix, and it is updated at each scan cycle by the elapsed time from the last scanning cycle.

The history file keeps track of actual performance on the shop floor. It keeps the following data:

Sequence number
Resource number
Product, item and operation
Start time
Finished time

The *sequence cycle loop* starts at the beginning of the session and whenever processing an operation is complete. At this point the resource becomes idle and a decision has to be made on the next assignment. The s*equence cycle time* is the elapsed time between present time and the previous sequence cycle loop. The time is retrieved from a running clock that starts at the beginning of the scheduling process and advances by the working time.

If the counter is zero it means that it was idle in the last scanning cycle, and will be treated as such.

If the counter is not zero, the sequence cycle time is deducted from the resource counter. If the result becomes zero it means that the process of the present operation is finished. In this case the priority field (PR) of this operation is marked by X, and the priorities of all operations depending on this operation number are changed to 00.

Automatically the next operation on that item becomes free and gets priority in processing, if it is economical to do so. This means that this resource is the "BEST" for this operation or that its processing time/cost minus a transfer penalty is equivalent to or lower than the "BEST" time of that operation. The operation is allocated to that resource, its resource status file is updated and its counter is set to the new operating time.

If the list is empty a "look ahead" feature is used to determine the "waiting time" for a best operation to become "free". This search is done by scanning the idle resource column for a free operation. When such an operation is encountered, (if it is not the best for that resource) the BEST field of this row indicates which resource is the best for that operation. The entry in the field ***resource counter*** of the ***resource status file*** indicates the waiting time of that resource.

2.5.2 Example

A list is compiled from the capacity planning module as displayed by matrices in Tables 4.3–4.6; the items planned to be processed in the first six periods are to be released to the shop floor for processing.

The selected items are: 1008 903 401 701 405 907 1004 706.

A matrix of all selected items including order quantities is presented in Table 4.11.

2 The Planning Steps

Table 4.11 Matrix of open jobs

R#7	R#8	R#9	#10	#11	#12	16	17	18	19	20
2.86	3.12	99	99	99	99	0	1,008	7	20	
2.03	2.33	2.57	99	99	99	0	903	7	49	
99	99	99	7.8	6.6	12.2	0	401	11	70	
99	99	99	7.8	6.6	12.2	0	702	11	53	58
99	99	99	7.8	6.6	12.2	0	405	11	20	
2.03	2.33	2.57	99	99	99	0	907	7	20	
2.86	3.12	99	99	99	99	0	1,004	7	52	
99	99	99	7.8	6.6	12.2	0	706	11	5	0
1.91	1.85	99	99	99	99	0	903	8	49	
1.45	1.72	1.96	99	99	99	0	803	7	69	

Column 16 holds operation status code Zero (0) means free operation
Column 17 holds part name and code
Column 18 holds the number of the *best resource* for the job
Column 19 holds quantity (see Table 4.11)
Column 20 holds modified combined quantity

- **Combine jobs rules:** Jobs of different orders might be combined in a single batch if they are the same item of different orders and: The processing time of the batch is less than (let say) 60 min
- If they are free operation at the same time cycle
- Similar single items will handle independently but will have priority to be loaded in union with one another

In the present example:

Item 707 with its five units is combined with item 703 and its quantity is increased to 58 units.
Item 907 with quantity of 20 units is below the minimum time and it is combined with item 903 with total quantity of 49 + 20 = 69 units.

Such a check is made whenever new items are released for processing i.e. become free operations.

Start loading. The clock is set at zero and a sequence cycle loop starts.
R#7 is the first resource that finds a free operation—item 903–1.
Its processing time is 69 × 2.03 + 30 (minutes handling time) = 170 min.
It is recorded in the History file and resource counter.
R#11 finds item 401 free. Processing time is 462 min it is loaded and recorded. (see Tables 4.12 and 4.13).

If a best resource cannot find an operation, a second round is initiated looking for alternatives.

For item 702 the best resource is R#11—if it is busy, second best is R#10 which we will assume to be free. This increases processing time by 70 min but saves waiting time of at least 170 min T = 58 × 7.8 + 30 = 482 .The system decides to use the alternative and records it.

Table 4.12 Resource counter—status

#R No.	Item no.	Start time	Time left initial	Time left initial	Remaining time
7	903–1	0.0	170	0	0
8	1,004	0.0	192	22	0
9					
10	702	0.0	487	317	295
11	401	0.0	462	292	270

Table 4.13 History file

No.	Item	R#	Start time	Finish time min
1	903–1	7	00.0	170
2	401	11	0,000	462
3	702	10	0.0	487
4	1,004	8	0.00	192
5	903–2	7	170	332
6	1,008	8	192	285
7	803–1	7	332	448
8	803–2	7	448	534
9	405	11	462	624
10	301–1	1	462	750
11	602	2	487	700
12	305	6	624	899
13	602+5	2	700	1,000
14	301	1	750	1,263
14	205	6	899	1,114
15	502	3	1,000	1,943
16	201	1	1,263	2,081

Item 1004's best resource is R#7—it is busy, second best is R#8 which is free. It increases processing time by 12 min but saves waiting time of at least 170 min T = 52 × 2,86 +30 = 192. The system decides to use the alternative and records it.

All free operations were loaded. Therefore, the second sequence cycle loop will start when the time left (Table 4.12) is minimum, in this case the clock time is 170 when operation 903–1 is finished.

The remaining time of all resources is reduced by 170, as can be seen in Table 4.12. This sets R#7 to zero remaining time and thus makes it a free resource.

Item 903–1 finishes released item 903–2 (second operation) and makes it a free operation.

R#7 is ready to load. The processing time is:

T = 69 × 1.91 + 30 = 162. Thus it will be finished at 170 + 162 = 332.

There are no additional free operations. The third sequence cycle loop will start when the time left (Table 4.12) is minimum, in this case the clock time is 22 when operation 1004 is finished.

The remaining time for all resources is reduced by 22, as can be seen in Table 4.12. This sets R#8 to zero remaining time and thus makes it a free resource and enables item 1008 to be loaded.

This process keeps on going till all open orders are processed. The dynamic changes cannot be shown in Table 4.12, therefore only the first movements are presented. The full dynamics can be seen on the History file Table 4.13. The simplified example we have presented may be modified, if needed to incorporate scheduling techniques such as split, overlap etc.

It took 2,081 min to process all orders. 2081:120 = 17.34 two-hour periods.

For ease of perception the data of Table 4.13 may be presented in a similar format as used before which used 2 h (120 min) periods.

Chapter 5
Product Specifications and Design

Abstract Product specification as well as product design are innovative tasks and depend on designer creativity. They are responsible for the success of the company's products, therefore they are entitled to perform this task as they find fit.

A variety of product specification methods are used by different designers and for different purposes. Some of these methods are presented here as representative of the nature of the task. Individual designers may use these examples in their own revised versions.

1 Introdution

A product has to seduce the customer with its options and appearance. To arrive at saleable product specifications, all company departments should be involved in the manufacturing cycle and the product must be compared to similar products all over the world.

Product design and process planning are the two most important tasks of the manufacturing process. They incur over 80% of the processing cost and consume 30% of the lead time. These two are interrelated and affect one another; therefore they should work together in a concurrent method.

Product specification as well as product design are innovative tasks and depend on designer creativity.

1.1 Product Design: Engineering Design

Engineering is the application of scientific principles and empirical knowledge in the creation of an appliance or machine to serve a given task. This given task may be defined either by a customer with a specific need or by management in its search for solutions to perceived problems. The designer is a problem solver who applies knowledge from such fields as physics, mathematics, pneumatics, electronics, metallurgy, strength of materials, dynamics, magnetic, acoustics and others, in order to find the solution to the given problem, namely, a new or improved product.

There is no single solution to a design problem, but rather a variety of possible solutions that offer a broad selection of optima. The solutions can come from different fields of engineering and apply a variety of concepts. The driving power in a machine, for example, can be electric, hydraulic, pneumatic, mechanical transmission, or an internal combustion engine. From this set of solutions, it is the designer's choice as to which one is the desired optimum. In making this choice, however, the designer is bound by constraint conditions that arise from physical laws, the limits of available resources, the time factor, company procedures, government regulations, and morality.

The design is iterative. The designer continually reexamines previous decisions in the light of new information gleaned as the design progresses. New, random ideas and concepts are applied until satisfactory results are obtained.

The designer faces the problem of predicting the performance of the design. There are many uncertainties, since not every characteristic can be computed on the basis of theoretical scientific knowledge or backed by practical experience. Nevertheless, the designer must make decisions, take responsibility, and hope to achieve an acceptable solution.

Successful designers must be knowledgeable in the following areas:

- Engineering science
- Engineering processes
- Engineering materials
- Engineering costs
- Standard components on the market
- Company standards and design policy
- Competitors' product design

They must also be skillful at:

- Problem and task definition.
- Data collection, literature survey, and application.
- Mathematical modeling and manipulation.
- Combining previously unrelated bodies of knowledge to create new combinations.
- Thinking in three dimensions.
- Developing ideas by means of sketches and drawings.
- Anticipating problems in a new design concept, and estimating their ease of solution.
- Making decisions under the constraints of incomplete information and conflicting requirements.
- Communicating to all levels.

1 Introduction

Thus a successful designer must have the following attributes:

- Curiosity
- Creativity
- Flexibility
- Open mindedness
- Patience,
- Judgment

There are two distinct phases in design work:

1. Design of the basic concept of the solution.
    In this phase designers employ their creativity. They are able to let their imaginations run free and come up with any wild idea. The more astonishing the ideas, the better the designer.
2. Decision and solutions.
    In this phase designers employ patience and technical know-how. They are constrained by rules, procedures, mathematical equations, and standard communication techniques.

It is not easy to switch one's state of mind and work on both phases simultaneously: however it is essential. Frequently, the difference between good and bad design resides in a lack of attention to details rather than in the basic concept. Details are frequently left to junior designers or draftsmen who are not fully aware of the problem and its solution. However, the process planner and the production engineers are obliged to accept these drawings without question. The general view is that designers (of the basic concepts) are born rather than made. Design is a highly individualistic, intuitive process. It is very rare that designers are able to describe how and why they have chosen a particular solution. Nevertheless, good designers try to convey this subjective process objectively by following a certain pattern. This pattern represents a general problem-solving technique rather than a solution to a particular problem.

1.2 Design Goals: Task Specifications

The designer's work must always be directed toward a goal. This goal is usually stated in general non-engineering terms without any implication as to the means to be adopted to achieve it.

It is important that the designer does not rush into solving the problem as stated in the goal. The purpose of the task must first be understood, and then must be converted into a set of quantitative engineering specifications. For example, if the goal is to design a conveyor belt, it should be realized that the purpose of the endeavor is to move items from one place to another. The conveyor belt is only one possible solution. Another possibility to be considered is rearrangement of the shop-floor layout in order to eliminate the need to move items.

The goal in designing an air-conditioning unit is to create comfortable conditions of temperature and humidity. If the problem is initially stated in broad, general terms, more possible solutions will be considered, thus enabling better solutions to be found.

The second stage is to transform the general terms used in the task specifications into values. This will be done by collecting information and by computations. The term "comfortable conditions" used in the air conditioning example must be converted into a statement of the form "room temperature of 22°C and relative humidity of 50%." Such factors as room size, the normal temperature in the area, time required to reach the desired conditions, the wall sizes and locations, and the number of people in the room must be specified. In addition, the amount of heat transfer and the air flow must be computed in order to reach a good engineering task specification. The engineering task specification does not worry about air-conditioning; it concerns itself with specified values of heat transfer.

The desired conditions stated above represent the primary objective to be met. However, there are many desirable but not mandatory secondary objectives for the designer to consider. In the above example, these would include minimizing the size, noise level, and cost of the air-conditioning system.

The designer must specify all of the secondary objectives before embarking on the evaluation and selection of the solution. However, whereas all solutions must fulfill the primary objective, they need not achieve all of the secondary ones. In fact, they are usually unable to do so, since many of the secondary objectives conflict with each other. Thus the extent to which the secondary objectives are fulfilled will provide the scale by which the solutions are evaluated. Consequently, it constitutes the difference between good and bad design. In order to be as objective as possible, the designer must rate the relative importance of these secondary objectives before selecting a particular solution. These ratings will later be used to evaluate ideas and decide on the best solution.

In general, secondary objectives might cover the following topics:

- Ease of operation
- Durability (product lifetime)
- Reliability (low maintenance)
- Efficiency (low operating cost)
- Safety
- Ease of maintenance
- Noise level
- Weight
- Floor space occupied
- Aesthetics
- Cost
- Ease of installation
- Ease of storage
- Ease of transportation
- Ease of production

1 Introduction

1.2.1 Generation of Solutions: Ideas

This is a pure innovation process depending on the creativity of the individual. However, few people actually make use of their full imagination potential. Use of the following techniques can aid in the realization of potential creativity:

- **Awareness that there are many approaches to solving a problem;** Care should be taken to avoid a predisposition to particular methods or ways of thinking.
- **Brainstorming;** Criticism and the fear of criticism are known to inhibit creative thinking. The idea generation stage should be separate from the analysis stage. Creative thinking, together with the multiplicity and free flow of ideas should be encouraged; in other words, quantity is important. Evaluation is not permitted at this stage.
- **Talking to people about problems;** The simple task of explaining a problem and its difficulties, even to an inattentive listener, can provide a fresh stimulus that allows the problem to be viewed in a new light.
- **Inversion;** Invert the problem. Let moving parts become stationary and stationary parts put in motion. Turn inputs into outputs. View it sideways, turn it upside down.
- **Analogy;** Think of solutions to similar problems in nature, literature, science fiction, and so on.
- **Systematic search ideas;** Break the problem into sub-problems and concentrate on ideas for each stage separately (from top downward)
- **Hypothesis;** Overcome difficulties by supposing that they have been solved and carry on the ideas from this starting point. Imagine situations, hoping that if they generate new ideas, a practical solution will follow.

This is a pure analytic process. The designer must apply one solution to the problem at hand. A good designer generates many solutions, but is aware that the best solution possible has not yet been found. However, there are time limits, and at this point the designer must select the best available solution and continue to develop it. Thus solution selection is a decision process.

All alternative solutions must achieve the primary objective. Therefore, the overall performance of the alternative solutions is evaluated according to the extent to which the secondary objectives are fulfilled. This creates a problem, since each secondary objective is measured with respect to a different scale and in terms of different dimensions.

The dimensionless decision matrix is one useful tool that the designer can employ to assist in decision making. Secondary objectives that have similar dimensions (e.g., minimum cost, operation expenses, and maintenance expenses) must be treated by a conventional mathematical technique. The use of the dimensionless decision matrix in such cases will result in biased logical decision.

Table 5.1 shows the dimensionless matrix; it consists of the following:

Secondary objective column. This column lists the secondary objectives that the designer wishes to use in the evaluation.

Table 5.1 Dimensionless decision table

Secondary Object. (i)	Weight	Alternative (j)							
		J = 1		J = 2		J = 3		J = 4	
		Rating	Value	Rating	Value	Rating	Value	Rating	Value
1	W_1	R_{11}	W_1*R_{11}	R_{21}	W_1*R_{21}	R_{31}	W_1*R_{31}	R_{41}	W_1*R_{31}
2	W_2	R_{12}	W_2*R_{12}	R_{22}	W_2*R_{22}	R_{32}	W_2*R_{32}	R_{42}	W_2*R_{32}
3	W_3	R_{13}	W_3*R_{13}	R_{23}	W_3*R_{23}	R_{33}	W_3*R_{33}	R_{43}	W_3*R_{33}
4	W_4	R_{14}	W_4*R_{14}	R_{24}	W_4*R_{24}	R_{314}	W_4*R_{14}	R_{44}	W_4*R_{14}
n	W_n	R_{1n}		R_{2n}		R_{3n}		R_{4n}	
Sub total ST_j			ST_1		ST_2		ST_3		
Certainty			C_1		C_2		C_3		C_4
Total			$P_1 = C_1*ST_1$		$P_2 = C_2*ST_2$		$P_3 = C_3*ST_3$		P_4

Weight column. This column lists the relative importance of the secondary objective. The least important secondary objective is given a value of 1. While the others will have values greater than 1, the specific value depends on their importance. The number of secondary objectives should be considered when assigning the weight.

The assignment of weights is a subjective task. However, designers who have to reach a decision must bear in mind that their decisions and designs will be judged by management and, in the end, by product marketability and customer satisfaction.

Alternative solution columns. These columns list the alternatives remaining after the rough evaluation. Some alternatives, especially if the brainstorming technique has been used, can be deleted simply by making a rough comparison with the other ideas.

Certainty row. The entries in this row are expressed as a coefficient that is less than 1. The certainty coefficient permits the inclusion of solutions that the designer is not familiar with, such as those that have appeared in the literature, those that have been used in other fields and problems, or those that appear as theoretically feasible. A good designer should be open-minded and consider unconventional solutions that have never been tried in practice. However, designers are not scientists, but engineers, and as such they will be judged not by their ideas, but by the results they achieve. Thus to be on the safe side, designers will stick to sure and familiar solutions. The certainty coefficient represents the amount of chance they are willing to take. Its value (expressed in percent) defines the advantage of the uncertain solution over the best solution in the dimensionless scale; hence, it is used to lower the overall rating received by the uncertain solution. For example, a laser beam provides one potential alternative metal-removing technique. The designer may be familiar with the literature on lasers, but has never used it in practice. This is no reason to disqualify it as an alternative; however, since it is the designer's responsibility to develop a successful design, unfamiliar alternatives must be regarded as somewhat risky. Thus the designer might decide that only if the laser alternative is superior to the other alternatives by more than 30% will taking the responsibility of using this idea be worth the risk. In this case, the designer will be using a certainty coefficient of $100\% - 30\% = 70\%$ (or 0.7), In order to maintain objectivity, the weight and certainty coefficients must be assigned as the first steps in solving the dimensionless decision matrix.

Rating column. The entries in this column will be graded on a dimensionless scale of 1–10, where 10 is assigned to the alternative that best satisfies the secondary objective under consideration; the other alternatives will be rated relative to this best alternative in a decreasing scale. More than one alternative can be assigned the same value (including the value 10); however, at least one alternative for each secondary objective considered must have the rating 10. Designers must depend on their judgment and estimates, since many details are as yet unknown. It is their job, and they will be evaluated on the basis of their achievements and decisions. They might choose the alternative they prefer or "feel" to be the best.

The dimensionless decision matrix is their tool, and they can use it any way they like; it might assist them in selling their ideas to management, but above all it helps them in their quest for a good and objective decision. Designers can also use the advice and judgment of others in the rating of the alternatives if they so desire. Dimensionless secondary objectives, such as aesthetics, will be rated by intuition. Dimensionless secondary objectives, such as minimum weight and cost, can be rated by intuition or by estimated values. If the secondary objective is minimum weight, the estimated weight can be used in determining the rating. Estimates all weight alternatives and assign the value of 10 to the alternatives with the minimum weight.

Since the above calculation is only an estimate, whole numbers may be used and the ratings computed and rounded as desired.

The selected alternative is the one for which the Pj expression has a minimum value.

1.3 Product Specifications Methods

Product specification as well as product design are innovative tasks and depend on designer creativity. The designer alone is responsible for the success of the company products, therefore he or she is entitled to perform this task as seen fit.

Several product specification methods are used by different designers and for different purposes. Some of these methods are presented here as examples. The designer may, if so desired, adapt or revise any method.

1.3.1 Gate List Method

The gate list is used to gather characteristics for the new product and serve as a project management tool in to the controller and the follow up of the project. It is composed of several gates, each one representing a major milestone in the development project. Each gate is then divided into internal milestones. The objective of the gate system is not to design or make decisions, but rather to control the advancement of the project and make sure that nothing was forgotten or neglected during the course of the development. Each project manager may set up individual major milestones.

In our example the project manager decided to use the following gates:

- Gate 1: Task specification
- Gate 2: Project contract
- Gate 3: Concept design
- Gate 4: Detail design
- Gate 5: Approval of design
- Gate 6: Final contract

1 Introduction

The objective of gate 1 is to make sure that the project definition is clear and complete, and that preliminary study was made before rushing into development. This gate was divided into four milestones:

- Preliminary conditions
- Initial analysis of requirements
- Initial solution search
- Project definition

The milestones and the subjects in each milestone are specific to a particular line of business, a master product. The subjects are general, and might not be applied to each project. The system does not propose solutions but rather indicates that a certain topic is of importance, an option or not applicable. The gate files refer to general technology, that which is needed to be specified or the available default values.

As can be seen, the first step in the project is the one which calls for studying the preliminary conditions, such as:

1. Is it a new project?
2. Is it an improvement of an existing one? Why? Was it changed?
3. What is the difference between the requirements and the Standard product?
4. What is the potential market?
5. What are the expected quantities?
   . . . .
   N. Etc

Next the initial analysis of requirements, understanding the secondary objectives of the product, such as any special requirements for:

1. Size
2. Weight
3. Volume
4. Mechanical strength, etc.
5. Assembly
6. Economic, cost, etc.
7. Heat dissipation
8. Fire resistance
9. Acoustics
10. Cost
11. Floor space
12. Recycling
13. Ease of maintenance, etc.

The initial conditions search exists to make sure that the engineer does not rush to develop the product. It is one of the group technology concepts, that "One solution may be used by many problems", or "Do not invent the wheel over and over again". The initial solution search covers topics such as:

1. Did you make a simulation to check the validity of the requirements?
2. Did you translate the requirements to workable data?

3. Did you isolate problematic requirements?
4. Did you perform experiments to validate possibilities?
5. Did you discuss with the customer the possibility of relaxing requirements?
6. Did you consider alternative designs?
7. Did you consider different methods of assembly?
8. Did you perform experiments to validate possibilities?

The project definition covers topics such as:

1. Did you check solutions against requirements?
2. Did you consult with members of other disciplines?
3. Did you consider alternatives?
4. Did you consider economic aspects of the product design?
5. Did you consider assembly problems?
6. Did you consider the maintenance problem?

Again, the gate list method does not propose solutions, but rather draws the attention of the developer to common possible solutions. Everything requiring the user's input has to be included so that the user can indicate yes/no to any possible solution, indicating that it was considered. The designers' decisions recorded in the gate file dialog session will create the required technical data file, in a similar style as the file for existing products.

1.3.2 Check List Method

Product definition is based on the collected data concerning products made in the past. Some aspects of the specifications are hidden by the decisions made, without being explained, but rather are incorporated in the design. Most of such specifications are of secondary importance, but they enrich the product and make it comparable to other products of the same field. In order to consider such specifications the checklist file is used. The use of the checklist serves as a remainder to draw the attention of the user to many topics. It allows the user either to specify that the topic is of no relevance to the specific product, or to enter additional requirements to the product specifications. A sample of a checklist file follows.

Did you consider the following items??

Check	Item
_____	Ease of operation
_____	Durability (product life)
_____	Reliability (low maintenance)
_____	Efficiency (low operation cost)
_____	Safety
_____	Ease of maintenance
_____	Noise level
_____	Weight

1   Introduction

- Floor space
- Aesthetics
- Cost
- Ease of installation
- Ease of storage
- Ease of transportation
- Ease of production
- Ease of handling and installation
- Ease of operation
- Stability
- Sensitivity
- Compatibility with its environment
- Ease of maintenance
- Durability long service life
- Reliability, low maintenance cost and short down time
- Efficiency low operation cost
- Height
- Volume, plan area, front area
- Flexibility
- Design for Production
- Ease of production
- Use of available resources
- Use of standard parts and methods
- Reduction of rejects and scrap parts and material
- Design for Distribution
- Suitability for storage
- Suitability for display
- Design for retirement of product
- Matching of physical life and service life
- Replacement ability of:
- complete product
- short-lived components
- Recovery of reusable material and long-lived components

1.3.3   Master Product Design

The basic concept behind this system is that each manufacturing company is in a specific line of products or business. A line of products usually has many common features. For example:

- Radar systems have a basis in electronics, servo system, power supplies etc.
- Communication products have a power supply, cabins, printed boards, antennas etc.

- Optical products have lenses, lens holders, polishing techniques, electronics etc.
- Rocket engines have nozzles, isolation components, propellants, guidance systems, sensors, electronics etc.

As one may notice, even in different lines of products there are several common sub-assemblies, not to mention in different models of the same product. Each has its own characteristics and requirements, yet there is something in common. The proposal is to study the products in order to come up with a "MASTER PRODUCT".

A "master product" is defined as a schematic block diagram of a product, where each block of the diagram represents its objective and includes alternatives, availability and cost information together with technical specifications.

The objective of the product definition module is to assist and guide the product specifier in defining a product that will meet all product objectives, drawing his or her attention to the effect of a decision on product cost and lead time.

The concept of master product was evolved and tested in an electronics company. The test results showed that product specifications that used to involve several months of negotiations with costumers can be done in a few days. Product design time was reduced to few weeks. Development cost was reduced to 10% of its original value and inventory showed remarkable savings.

The philosophy behind this proposal is that product definition is an innovation process and, as such, must leave freedom and judgment to the individual that performs this task. The role of the computer, if any, is secondary. The backbone of the system is technology and not computers. The system draws the attention of the user to the meaning of his or her decisions and, in some cases, to propose alternatives. However, the final decision is left to the user.

The product specification has to seduce the customer, with its options and appearance. To arrive at such specifications, several stages of the manufacturing cycle should be involved, and a comparison to similar products, produced all over the world, must be made (Benchmarking, One of a Kind, World Class Manufacturing, etc.). Furthermore, to arrive at a product design that will result in low cost, ease of manufacturing and ease of assembly design techniques such as DFM, DFA, Concurrent Engineering must be considered.

The type of activities in which a company is engaged is displayed in the product definition stage.

1. A company may manufacture to order. In such cases the product is defined by the customer. However, it will be a good idea to consult with the manufacturer on the design of the production order to reduce manufacturing costs.
2. A company manufactures a line of existing products. In such cases, it is a good idea to keep track of possible design changes that might reduce manufacturing costs, increase product appeal to customers, and introduce options and new models of the same product.
3. The company would like to enter a new field of activities.

The master product methodology supply answers for all three types of activities.

1 Introduction

The Master product design is the main program that controls and navigates the session. The general data of the session, such as title, name of user etc. are being recorded in the specific product design files.

The session proceeds by presenting inquiries to the user, asking for specification of intentions and wishes. Usually the user comes from management and may not be familiar with the terms used by the system. Therefore, the user may ask for "help" and get explanations and clarifications of the topic in question.

The first step in specifying a product is to understand, and have the relevant data of, the purpose of the task. Only then can the product specifications be formalized. The logic at this stage is that it is a natural tendency for the one who specifies product characteristics to aim for the best, and rightfully so. However, the specifier is not always aware of the costs and manufacturing implications.

In many cases, reducing the specified values by as little as 5% may result in cost reduction of more than 60%. We assume that the product specifier may change the specifications after becoming aware of this effect.

Product specifications must be set in specific terms and not in general terms such as nice, light weight, accurate. It has to set values and tolerances to each term, (like tolerances in a drawing). For example when the product is a weighing scale, the accuracy must be specified. If it is a bathroom scale it is nice to have a scale with an accuracy of +/− 1 g. It can be done. However, it calls for costly accurate sensors and support circuits. If the accuracy is specified as +/−100 g, a simple sensor and a low cost product will result. However, if the scale is used in a factory for packing candy units of 100 g, the accuracy must be +/− 1 g or less.

Messages drawing the attention of the user to such effects will be presented whenever the response to the inquiry is non-standard or out of reasonable values. To determine if and when to post a message a ***Technical Data file*** will be used.

Auxiliary files and databases are used to assist the product specifier and product designer, in the fields that are out of the specifier's expertise. These databases represent the interest of all other disciplines of the manufacturing process. Hence the user need not always summon a group meeting of experts from different disciplines for consultation; the same results can be achieved just by sitting at a working station. The sessions can be more efficient and objective, arrive at better decisions, and save time.

Technical Data

The technical data file contains data relevant to the master product, i.e. it is general data for a family of products. With the guidelines of the master product design session, the customer converts the master product to a specific product that suits his or her needs. The "customer" is referred to as an external one, in case of custom-made products, or the marketing or sales department in designing commodities products.

Any product is used to fill a certain need. To achieve these objectives, certain functions must be available, and have to be specified accordingly. They will be called primary objectives. However, there are many features that do not contribute to the main objectives of the product but come to serve as supporters. They will be called secondary objectives.

For example, in designing a scale, the measuring unit and the display units are essential for the product's performance. But the method by which the display unit is presented and its shape is irrelevant to the product performance. It is a marketing feature and not a performance feature. To carry it to the extreme, a factory scale does not necessarily need a display at all.

Marketing must be consulted in making such a decision. In many cases, the secondary objectives might contribute to the cost and lead time more than the main objectives. The secondary objectives of different disciplines might contradict each other. Mainly because the discipline experts are not experts in other fields and they may not be aware of the implications of their specified objective on the manufacturing process. A compromise between the different objectives is needed. Usually it takes several meetings, with fierce argument till a decision is taken. In many cases it is a biased decision. Bias as personal favoritism along with the gift of speech, is a dominant factor in the decision making process.

To consider these secondary objectives, a note file will be used. An example of data in this file, for the case of a weighing-scale product, might be as follows:

1. The weights range from 2 to 50 Kg.
2. The platform should be able to hold parts of 1.5×1.5 m.
3. The scale is to be mounted on a floor, near a wall.
4. The maintenance should only be done from the front side.
5. The display should be digital.
6. The display should be readable from a distance of 5 m.
7. The display should present set of tolerances to be set manually.
8. Setting the range of tolerances should be made by hand without special tools.
9. An alarm should sound when exceeding the upper tolerance.
10. The maximum cost should be $1,234.
11. Assembly should be done at the client's location.
12. The scale should withstand sock load.

Product Specifications Example

Master product design is not a universal system; it is aimed to assist a specific plant that is in a specific line of business. Normally we assume that the plant is in business for quite some time and have a line of products. The steps of constructing the system are as follows:

1 Introduction 115

The first step in installing the master product design system is to study the existing product. A block diagram of each product will be drawn. The block diagram should not be a detailed one, but rather present the main module of the product. The block diagram should indicate the objective of each module, and the interrelationship of the different modules.

Comparing block diagrams of different products in the search for a family of products. A family of products is defined as group of products that have the same or similar block diagram. The block diagram of the individual products in the family need not have exactly the same block diagrams, but such that the same blocks appear in most product diagrams.

The family block diagram will be declared as a master block diagram, and list the available products that are included in the family. Figure 5.1 shows a family of products for security gates. There are many types of security gates (barriers), and each has advantages and disadvantages. However, all can be described by a single block diagram as shown in Fig. 5.2. The operation is as follows:

1. A signal is transmitted to open the gate.
2. The signal is interpreted and amplified to actuate the power source.
3. The movement of the power source drives the doorway mechanism and opens the door.
4. A feedback signal is transmitted to the signal interpreter and according to the program the door is closed.

The objectives of all products in the family are the same, however there is a difference in design among the individual products. The reasons for the difference are studied in order to understand the products. Such study might reveal that the purposes of the products are different, and that the products are being used in different environments, with different required characteristics. The study of the characteristics indicates the way to specify a product. The results of such study are entered in the technical data file. The object of this file is to assist and guide in forming the dialog session, and supply data for messages during the session.

Using the example of the security gates, it was found that these products are used in a civilian private apartment's buildings or homes, in public parking lots, or by the military. Each one of the different users calls for different characteristics of the gate. The difference might be in: the speed of operation, the security level that it provides, and in the ease of operation. Another important factor is the reliability of the doorway action. Following the above study a technical data file is constructed as shown in Table 5.2. The values are the range of the different products that the company made in the past.

The ***data collected*** can be used to form the product design dialog. The users' final responses to the product design dialog are stored in the specific product design file, and are regarded as the customer specifications for the ordered product.

Fig. 5.1 Family of security gates

1 Introduction

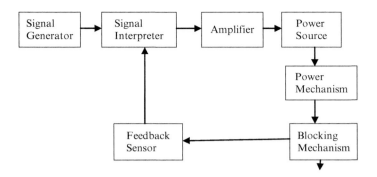

Fig. 5.2 Master design block diagram

Table 5.2 Example of technical data file study for security gates

User	Security level	Strength mKg	Freq.per hour	Speed seconds	Height meter	Length meter
Private mall	2–3	40–60	2–5	2–4	1–1.5	1–2
Private medium	2–4	60–80	4–7	2–4	1–1.5	1–2
Private large	2–4	60–80	5–10	1–3	1–2	1.5–3
Parking lot	4–5	80–100	40–50	0.5–1	1–2	2–4
Street blocking	3–4	70–200	2–5	2–4	0.3–.8	0.6–1.2
Office building	6–8	80–120	30–40	0.6–1.0	1–2	2–4
Military camp	8–9	400	40–60	1–2	1–2	2–4
Military hangar	8–10	300	–	–	–	–

Continuing with the example of security gates, the session to define product specifications proceeds as follows:

System inquiry	User response	System messages
State the purpose	Help	Private small
		Private medium
		Private large
		Parking lot
		Street blocking
		Office building
		Military camp
		Military hangar
		NONE of the above
	private medium	OK
State the required doorway length	3 m	Are you sure? The normal length for private medium doorways is 1–2 m
	I am sure leave it	
	3 m	OK
State the required doorway height	1 m	OK

(continued)

(continued)

System inquiry	User response	System messages
State the required doorway strength	1,500 Nm	Are you sure? The normal strength for private medium doorways is 60–80 Nm. Please note that the higher the strength the higher the cost. Cost estimation—direct ratio of required strength to the medium default value.
	Make it 100 Nm	OK
State the required operating speed	0.5 s	Are you sure? The normal speed for private medium doorways is 2–4 s. Please note that the higher the speed the higher the cost. Cost estimation—increase by 10% per reduction of 1 m/s
	make it 3 s	OK
State the required frequency of operation	20 times per hour	Are you sure? The normal frequency for private medium doorways is 4–7 times per hour. Please note that the higher the frequency the higher the cost. Cost estimation—increase by 10% per reduction of 1 m/s
	make it 5 times per hour	OK
State the requirement security level	6	Are you sure? The normal security level for private medium doorways is 2–4. Please note that the higher the security level the higher the cost. Cost estimation—increase by 150% per increase of 1 level from the default
	I want it to be level 6	OK
Please select the doorway style	Help	The available styles are: Swing barrier Sliding sideways barrier Slide upside barrier Cylindrical barrier Sharp teeth barrier NONE of the above
	Swing barrier	The swing barrier style cannot support level 6 security. The recommended style for level 6 security is the Sliding door. Do you wish to modify the style or the security level?
	Change style to Sliding sideways barrier	OK Usually a sliding barrier is of 2 m high. Would you like to modify the specified height of 1 m?
	Yes, change to 2 m	OK

1 Introduction

The output of the dialog session is stored in the specific product design, along with the master block diagram. For the dialog presented above, the specific product design is called for:

- A sliding sideways barrier doorway style
- Three meters long and 2 m high
- Resist forced opening of up to 1,000 Nm
- No maintenance will be required for up to 2 years when average usage is up to five times per hour and delay in opening the gate from arrival to the opening signal is up to 3 s
- Security level of 6

The product definition session as shown above, is based on the collected data concerning products made in the past. Some aspects of the specifications are hidden by the decisions made, without being explained, but rather are incorporated in the design. Most of such specifications are of secondary importance, but they enrich the product and make it comparable to other products of the same field.

1.4 Master Product Design System: Concept Design

In Sect. 1.3 the technical data file was constructed to assist in specifying product characteristics. The next step is concept design. The objective of the master product design system in this stage is, as before, to propose solutions, to draw the attention of the designer to the possibilities and characteristics of each solution. In order to supply such data, in a practical way, the group technology technique will be employed. The following steps are for each one of the block diagrams of the master product design, or for a group of blocks that are affected by one another. The process will be repeated for each one of the blocks in the block diagram.

The steps in constructing the technical data base are as follows:

The first step is to study the topic of each single block of the existing product. A list of all designs of that topic that were used is made, including data on their performance, cost, and characteristics. Reference to the detailed design is made including its bill-of-materials. The different designs will be categorized by their method of operation, and the parameters that are important to the basic product design.

For the example of the security gate, the first block is signal generator, therefore the operation characteristics will be: security level, ease of operation and cost. Table 5.3 shows such a list.

Each type of entry and principle of operation on the list is then analyzed, comparing its performance history to the similar design concepts that were used in other products. The date of the design is compared to find out if there were problems with the specific design. New components that have appeared on the market are added to the technical data file so they can be considered in design changes or new

Table 5.3 Technical data file—signal generator options

Signal generator	Notes
Push bottom	Low cost, simple to operate, No need for interpreter
	No security Cannot be operated from distance
	cost = 1; security = 1; ease = 1.2
Push bottom—locked by key	Same as before, increase security—reduces ease of operation.
	cost = 1.2; security = 2; easy = 1
Hand transmitter	Medium-low cost, needs simple interpreter, can be operated from distance,
	Security depends on number of transmitting codes
	Cost increase as the number of codes (frequencies) increased. Medium-low security
	cost = 2.2–5; security = 2–6; ease = 10
Hand transmitter with scrambler	Same as before, increase security and cost
	cost = 4–8; security = 4–8; ease = 10
Plastic coded card	Medium-low cost, needs simple interpreter, cannot be operated from distance,
	Medium-low security cost = 3; security = 3; ease = 2
Key pad	Medium cost, needs simple interpreter, cannot be operated from distance
	cost = 4; security = 5; ease = 2
Noise signal (voice activated device)	Low—medium to medium-high cost depends on the noise signal used. Simple number of horn blow:
	cost = 2; security = 5; ease = 10
	Music combination; cost = 7; security = 9; ease = 10
Voice control- personal voice recognition	Medium high cost. Needs an interpreter high security
	cost = 5–9; security = 7–10; ease = 10
Finger prints	Medium high cost, needs and expensive interpreter, cannot be operated from distance
	cost = 8; security = 10; ease = 2
Live guard at the gate	Manual operation. Needs only the blocking mechanism, all other functions may be operated manually
	cost = ?; security = 8–10; ease = 8–10

designs. The objective of this study is to eliminate making the same mistakes that were done in the past, and find the best design. During this study proposals are made for upgrading and improving existing products by standardization, i.e. using the same best solution in as many products as possible. Naturally, many design solutions are on the list, in addition to customer satisfaction data, and maintenance department remarks. Such data will be part of the technical data file.

The list includes all designs that were made in the past. However, a design might not have been used in the past but might be a better one. The system will use the gate method, the check list to search for such a design.

A file will include the notes of different disciplines regarding each concept of design. The list includes all designs that were made in the past.

Table 5.4 shows an example of the technical data file for a scale.

1 Introduction

Table 5.4 Sample of a technical data file: weighing mechanism

No.	Type	Principle of operation	Remarks	Cost
1	Deflection: Coil spring Leaf Spring Column	mechanical levers	Standard product in different sizes Low accuracy (± 60 g) low repeatability non linear at the ends	Very Low
		Potentiometer (linear or rotational)	Medium accuracy (30 g) Medium linearity Easy to connect to display Require power supply Simple electrical circuit	Low
		LVDT (Linear Variable Differential Transformer)	Very accurate (1 g) Require power supply Low power output Needs amplification	High
		Induction Coil & Capacitor	Very accurate (0.4 g) Require power supply Low power output Needs amplification Complicated electronic circuit	High+
2	Stress & Strain Bar	Strain gauge	Very accurate (0.1 g) Require power supply Low power output Needs amplification Complicated electronic circuit Mechanical simple	Very high
3	Load Cell	Purchased item (usually based on strain gauge)	Very accurate (0.5 g)	Expensive

A design is selected by comparing the product specifications as defined by the first stage, with the capabilities of each design solution, as appears in the appropriate technical data file. If several designs meet the requirements, secondary design objectives will be recorded and an analysis will be made as to the best compromise. Instead of making automatic decisions, the system displays a table of option design with grading according to the system strategy, and the customer makes the final decision. All the necessary data will be presented to the customer in order to make an intelligent decision, based on facts and real data.

The system considers the concept design and the specific product design as one system with different design objectives, but using exactly the same solution strategy. In this stage each block of the master product design will be treated as a separate design objective.

For example: the first block of the security gate is the signal generator. Hence the problem on hand is how to design a signal generator. Examining the specific product requirements indicates that a security level of 6 is required. A search is made through the signal generator technical data file in search for a design that will support the required security level of 6 or above.

The following concept solutions are the alternatives and the message to the designer might look like:

Option 1—Hand transmitter

> Medium-low cost, needs simple interpreter, can be operated from distance, Security depends on number of transmitting codes (frequencies). Cost increase as the number of codes increased. Medium-low security
> cost = 2.2–5; security = 2–6; ease = 10

This is the simplest and the least expensive design. The transmitter can be duplicated, allowing many people to operate the gate. It is very easy to operate, lightweight. It can be operated from a distance of 5–20 meters.

Option 2—Hand transmitter with scrambler

> Same as before, increase security and cost
> cost = 4–8; security = 4–8; ease = 10

Option 3—Noise base signal (Voice activated devices)

> Low-medium to medium-high cost depends on the noise signal used.
> Music combination: cost = 7; security = 9; ease = 10

Option 4—Voice recognition

> Medium high cost. Needs an interpreter high security
> cost = 5–9; security = 7–10; ease = 10

Only authorized (recognized) persons may operate the door. Each one has to teach the system the sound of his voice. Therefore it provides high security. However there might be problems with voice changes. The distance with which it may be operated is limited to 0.5–2 m.

1 Introduction 123

Option 5—Fingerprint checking

> Medium high cost, needs and expensive interpreter, cannot be operated from distance
> cost = 8; security = 10; ease = 2

Only authorized (recognized) persons may operate the door. Each one has to give the system a sample of his or her fingerprints. Therefore it provides high security, and can be operated by touch.

Option 6—Live guard at the gate

> Manual operation. Needs only the blocking mechanism, all other functions may be operated manually
> cost = ?; security = 8–10; ease = 8–10

Low initial cost, but run an expensive operating cost.

SYSTEM	The system recommends using option 1. Please confirm.
DESIGNER RESPONSE	I would like to use option 2
SYSTEM	OK

A similar session is conducted for each of the block diagrams.

For example, the power source session might include proposals for electric motors, hydraulic cylinders, pneumatic cylinders, linear electric motors, hydraulic motors, wire pulleys, etc. with information regarding the pros and cons of each option, and its implication on the power mechanism.

Examples of such data are as follows:

Hydraulic operated power source requires an auxiliary hydraulic system including pump, regulating valves, and an accumulator, which increases the cost of the system. A hydraulic power source has a better force- weight ratio than an electric motor. A hydraulic cylinder supplies linear motion while an electric motor gives rotary motion that should be translated into a linear motion. The translation, gear train, rack and pinion might increase the security.

1.5 *Master Design System: Detail Design*

In the previous section the concept design was determined. This decision will guide the designer in the detail design. The concept of performing the detail design is the same as the method used for the concept design and the product specifications. The difference will be in the content of the technical data file. In this stage the data will include references to existing drawing of similar detail design, and comments on their performance. A comparison of previous designs for similar objectives will result in a master detail design of the product. It will include the assembly drawing (product structure), and shape of each item. This serves two purposes.

In the initial stage of constructing the data file, the comparison, duplicate designs will probably be recognized. Many such duplicates are not due to the needs of the product, but perhaps because the designs were made by different designers or at different time periods. When such a phenomenon is encountered, an *upgrade of existing products* can be achieved by selecting the best design and using it for all other products.

Second, the best design may be studied in search of improvements. It is possible that, as time passes by, new technologies, materials, components that may be used in the design, but were not available at the time of the original design, will be discovered. Such design modifications may be introduced, if they provide economic benefits, or more important be regarded as a master design for future products. Using the gate method, the note file and the check list, will provide valuable tools for selecting the master detail assembly design (product structure).

This design is regarded as a master design, as it provides a general view of the product and its assembly, but, without exact dimensions and tolerances. Such details might differ for each product objective, and it is up to the designer to make these decisions. One should not forget that, although the system is heavily based on computer programs and databases, design is still an innovation process and the designer is the one who makes the decisions. The computer's role is to draw attention to certain features, to propose designs, to minimize the need for team meetings as much as possible, but not to make decisions.

The technical data files for this stage will include engineering handbook data such as data on materials and their specifications, local and international standards, tolerance tables, useful equations, screws, bolts, rivets etc. and other data that is relevant to the specific company. A call for HELP will assist the designer in selecting components. A list of available products is displayed, along with an explanation of the differences among the options and the advantages of and problems with each option. An additional call for help might provide information on different methods of connection and their effect on cross-section shapes.

For example: "bearings" will list the available products, such as: radial bearings; thrust bearings; slide bearings; ball bearings; roller bearings; and explain the difference, advantages and problems with each one of the motion-supporting devices (bearings). Similarly the attaching of parts HELP might provide the following information: method of connection: rivets; screw & bolt; weld (list different types); glue; adhesive; magnet; free slots (hold one direction—two directions). The effect of connection on cross section shapes, of each method, will be explained.

In addition to the above files and databanks, an inventory file and suppliers file is attached. Their purpose is to attempt to use standard products and available stock, thereby reducing dead stock if available and avoiding delivery delays. Often, there is not one best solution or material, but many solutions. The difference in cost and performance is not always significant, especially if the cost of dead stock is being taken into account.

1.5.1 Technical Drawings

The design decisions reached in the engineering design stage are transferred to the process planning and other manufacturing stages in the form of technical drawings. The technical drawings should include complete information on the geometry and associated data, such as: geometric shape of the parts and product, its dimensions, tolerances, geometric tolerances, surface finish, and the raw material.

Each one of these data if not carefully assigned might misrepresent the designer's intentions. Moreover, processing cost and time are functions of the dimension method. A close tolerance calls for extra process operations; a loose tolerance might cause parts rejections. This stage of design is not as glorious a one as the concept design, but its effect might be as damaging as a wrong concept, and should be treated very seriously. Some examples of hazardous traps that should be avoided are presented.

The roadmap matrix that might be used by the designer is used in evaluating processing cost and lead time of different design and dimensioning options.

Dimensioning and Tolerance

A part should be defined in such a way that, when assembled it will fulfill its technical functions and be tolerant so that it can be mounted in a subset of parts in properly dimensioned and a completely interchangeable manner.

To dimension parts that would assemble with each other, the dimensioning should originate at a *datum*. A datum is a theoretical ideal plane, line, or point. A datum is a physical feature of the part identified on the drawing by a datum feature symbol and corresponding datum feature triangle. These are then referred to by one or more datum feature references which indicates measurements should be made with respect to the corresponding datum feature and may be found in a datum reference frame. Datum is usually marked with a letter of the alphabet and placed in a box attached to the edge view of the surface.

A drawing may of course contain any unimportant details which have nothing to do with functioning and assembly. The dimensions for these need not originate at a datum.

In a given direction, a surface should be located by one and only one dimension. Much confusion can arise from violating this rule.

For example, consider the horizontal dimension of a part shown in Fig. 5.3. It includes three dimensions: A, B, and C.

A redundancy occurs when all three dimensions are given as:

$$A = 50; \ B = 30; \ C = 80.$$

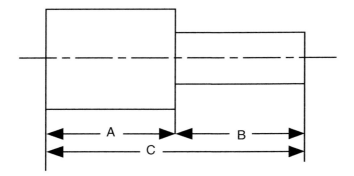

Fig. 5.3 A redundant dimension

The arithmetic is correct but due to variations in processing (tolerances) the part cannot meet the defined tolerances, which might be for example as:

$$A = 50 \pm 0.1; \ B = 30 \pm 0.1; \ C = 80 \pm 0.1.$$

A redundant dimension can be the cause of out-of tolerance parts.

The difficulty can be corrected by omitting one of the dimensions. The two dimensions that should be retained depend on manufacturing convenience or the functional requirements of the part. From the discussion above it is obvious that only sufficient dimensions should be placed on a drawing. Any additional dimensions will nearly always result in parts that meet the drawing but out of the specified tolerances.

To meet the functions of a part due to machine inaccuracies, any dimension on a drawing must be accompanied by tolerances. The stack-up tolerances are a function of the dimensioning method assigned by the designer. The basics of tolerance arithmetic are explained in the following examples:

Figure 5.4a shows a chain of four dimensions with their tolerances. One task is to define overall length of the part. The nominal length will be obviously:

$$L = A + B + C + D.$$

The maximum length will be:

$$A + a + B + b + C + c + D + d = A + B + C + D + (a + b + c + d).$$

The minimum length will be:

$$A - a + B - b + C - c + D - d = A + B + C + D - (a + b + c + d).$$

And the length tolerance will be:

$$l = a + b + c + d.$$

1 Introduction

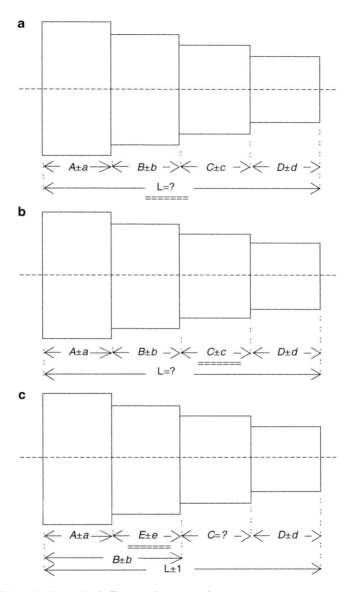

Fig. 5.4 Dimensioning method effect on tolerance stack-up

Figure 5.4b shows the total length with its tolerance (L ± l) as well as the tolerance of dimensions A, B, D. The problem is to define the tolerance of C.

The nominal dimension of C is:

$$C = L - (A + B + D).$$

The maximum length will be:

$$C = L - (A + B + D) + (l + a + b + d).$$

The minimum length will be:

$$C = L - (A + B + D) - (l + a + b + d).$$

And the C tolerance will be:

$$c = l + a + b + d.$$

The resulting dimension is therefore:

$$C \pm c(l + a + b + d).$$

These results show that whether the dimensions are added or subtracted, the resulting law of tolerance is as follows:

The interval tolerance of the result is equal to the sum of the tolerance of the components.

Figure 5.4c shows an example with the same four dimensions except that A and E are not dimensioned individually, but their sum is B. If B is tolerance as before, the tolerances of A and E have to be reduced (dimension A or E should be omitted). On the other hand, the tolerance of C will be reduced to $c = l + b + d$, assuming of course that the different tolerances are of the same magnitude as in the cases of (a) and (b).

Geometric Tolerances

All bodies are three dimensional, and on an engineering drawing a body is assumed to be placed in a system of three perfect smooth planes oriented exactly 90° to each other. However, perfect planes cannot be produced. The shape tolerances cannot guarantee that the part produced will meet the designer's intentions.

For example: At the top of Fig. 5.5a drawing of a straight shaft diameter of Ø150 ± 0.5 is shown. At the bottom of the figure the produced part is shown. The produced part meets the specified tolerance. At any cross section of the part along its length, the diameter will be Ø150 ± 0.5 however the center line is not a straight line but a curve. No indication on the drawing prevents such a curve. Furthermore, the shape must not be a perfect circle, as can be seen by Fig. 5.5. The circularity of the part must be within two circles, one of Ø150 and the other Ø150.5.

Another example is shown in Fig. 5.6. The drawing on the left shows the designer's intentions, and on the right side the produced part that meets the drawing specifications. The drawing does not specify that the two cylinders must be concentric.

1 Introduction

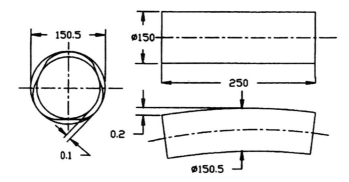

Fig. 5.5 Possible good part that meets diameter tolerance

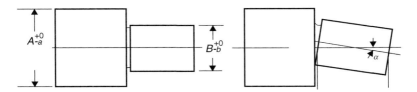

Fig. 5.6 Part that meets drawing specifications

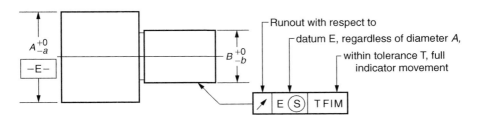

Fig. 5.7 Shaft drawing defining runout

Geometric tolerances come to enable the designer to specify more precisely his intentions.

Figure 5.7 demonstrates how to use geometric tolerance to remedy the problem presented in Fig. 5.6.

The geometric tolerances symbols are placed in a box adjacent to the dimension to which they refer. The box also gives the appropriate datum and the geometric tolerance for the feature.

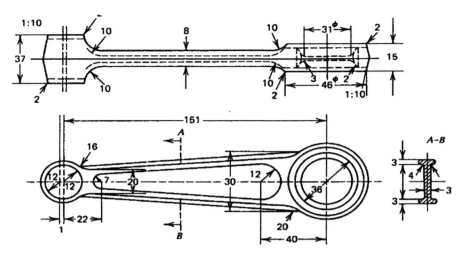

Fig. 5.8 Connecting rod

Design for Manufacturing

There is a saying on the shop floor that paper can tolerate anything. Which means that you may make any drawing you wish, but to be processed the drawing must follow some physical and natural rules. Just as a remainder two cases follows.

The connecting rod function is to transfer motion and torque between the crank shaft and the piston. This objective can be achieved by any design that keeps two holes with the correct diameter at a specific distance. A plain bar of, let us say $15 \times 30 \times 200$ mm, can be a good, easy, low-cost solution for a low quantity. For mass production it is a bad design. A better processing will be to produce the connecting rod by forging.

Forging is defined as the working of a piece of material into a desired shape by hammering or pressing, usually after heating to improve its plasticity. The material may be shaped by drawing out, which decreases the cross-sectional area and increases the length; by upsetting, which increases cross-sectional area and decreases length; or by squeezing in close impression dies.

Design parts for forging processing (as well as plastic molding) have some rules to follow: to keep material flow do not ask for sharp corners, allow a generous radius, try to keep wall thickness a unit, if thickness change is required make it gradual etc. Otherwise the rate of rejects will increase.

Therefore the design should follow such rules; a good design of a connecting rod is shown in Fig. 5.8. It follows the rule, keeps the material strength, low weight and functional.

Holes in plastic injection processing.

1 Introduction

Fig. 5.9 Positioning of holes

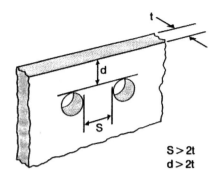

By introducing cores, a variety of hole shapes and sizes through the walls of a component is possible. A core that forms a hole however tends to limit natural material flow and results in an area of relatively high residual stresses around the hole. Round holes are generally less susceptible to this disadvantage, since the residual stresses tend to be evenly distributed.

The inclusion of any hole results in an interruption of the material flow around the core. Where the material rejoins, it forms what is known as a 'weld-line'. This is always weaker than the bulk of the material and may also be aesthetically undesirable. However, steps can be taken during both design and processing stages to minimize these problems.

To encourage the formation of a strong weld-line, the following points are recommended when specifying dimensions, geometry and position at the design stage. See Fig. 5.9.

- The shortest distance between the edges of any two holes or slots should be greater than twice the nominal wall thickness.
- When positioning a hole or slot near to the edge of a component, the shortest distance between the edges of the hole and component should exceed twice the nominal wall thickness.

1.6 *Summary*

The master product method is based on the fact that each manufacturing company is in a specific line of products or business. A line of products usually has many common features. Studying the different products and constructing a master block diagram and the auxiliary files, including roadmap system will result in an all-embracing manufacturing system.

Master product design is *not* a computerized system; it is a technological system, assisted by computers. One cannot have a standard, off the shelf product like this. It is unique to each plant. The supporting computer program is very simple to write. The file organization of the auxiliary files may be by keyword as referred to in the check list and a sequential search.

The main effort in constructing the master product design method is to collect engineering technical data, and market data. It is a huge job, but it is worthwhile. This study proved that within a few days an optimum product definition that gets the approval of all disciplines has been specified.

The benefits of the master product method are: shortening the product specifications and design from several months to several weeks, and getting a better product and design.

The plant where the system was organized saved over $9 million on a $15million project.

It has been realized that a computer system that assists in defining new product may be used for budget control, and can suggest options and alternatives in cases of changes in market demand and technology. Moreover, the computer program may be used by managers to follow up the progress of projects in the sense of time keeping, and more important in the sense of assuring that all relevant factors and procedures have been considered.

Chapter 6
Detail Design

Abstract Product design and process planning are the two most important tasks of the manufacturing process. They incur over 80% of the processing cost and account for 30% of the lead time. The designer's main effort is devoted to concept design, supported by draftsmen who fill in minor details.

Our intention in this section is to draw attention to minor design details that might reduce processing time.

1 Introduction

In the mechanical and electrical engineering industry, about one half of the working force is employed in assembly. Costs and manufacturing times of many products are determined, to a large extent, by the assembly process. It is thus clear that very great importance is to be attributed to correct design.

The first objective of assembly planning is to assist the designer in considering design for assembly in an organized manner. Each aspect of the activity should be considered in a logical sequence, so that the implication of decisions which are made are both known and consistent with decisions which might have been made had someone else been carrying out the study.

There are a number of reasons for considering assembly planning but generally one is seeking a reduction in operation cost. The other main reasons are:

- The long lead time in the assembly department with a high product value.
- High personnel input and hence high labor costs.
- Relatively large proportion of activities which cannot be counted as part of the actual assembly process.
- To increase output of an existing product.
- To improve consistency of quality and reliability.
- To seek solution for small batches which do not justify hard automation.

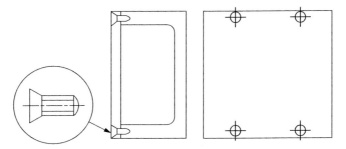

Fig. 6.1 Design of a box and a cover

- To reduce the problem of labor turnover, scarcity, or fluctuation in output due to minor labor disruption or absenteeism.

Engineering design is a specialized process of problem solving. Although it has its own peculiar way, suited to a technological pattern, its process resembles that of problem solving in general. There is always more than one solution to a problem, and probably there is not one solution that is the "best". There are many factors that the designer should consider, such as:

- Design for operational economy
- Design for ease of assembly
- Design for ease of processing
- Design for functionality
- Design for reliability
- Design for maintenance
- Design for safety
- Design for convenience of use

Several of these parameters conflict with one another and a compromise must be made. The designer has to do his or her best in the limited time assigned for the problem. In this section we consider mainly the first three.

The designer might argue that, instinctively, he or she always has in mind how a product is to be processed and assembled and that traditional design rules and common sense are sufficient. This may be, to some extent true, but surveys of a wide range of products show that in more than 90% of all products designed for processing and assembly, improvements could be made without affecting product functionality. Moreover, if a group of designers were given the same problem, they would inevitably produce a variety of designs which they all consider to be easy for processing and assembly.

A simple example of possible design improvement is demonstrated in Fig. 6.1. The designer's task was to design a box to contain a certain amount of volume and a cover. The design in the figure meets the product objective. However from an assembly point of view it poses several difficulties. The assembly task includes two operations: to position the cover in place; and to fasten it by screws. As the cover is

1 Introduction

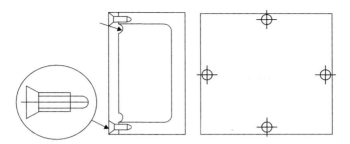

Fig. 6.2 Improved design of a box and a cover

designed as a flat square, the positioning operation is just to lay it down on the box and adjust its sides to the box sides. In addition the screw holes have to coincide with the threaded holes on the box.

The designer decided that four screws are needed to hold the cover and secure the content of the box in place. Therefore he put two screws in opposite sides of the square. By this decision, the positioning of the cover has to be in two stages; first put the cover on the box, then make sure that the holes coincide; if not the cover has to be turned by 90°, which calls for the assembler's attention and visual inspection.

A better design would be to space the holes in a symmetrical way, which means that if the sides of the cover coincide with those of the box, the holes of the cover and the box will match. This design improvement still calls for the assembler's attention in positioning the cover.

To eliminate the assembler's attention in positioning the cover, a dent in the cover might assure that the dent falls in place and the cover is in the right position, ready to fasten the screws.

A further improvement is to use self starting screws. An improved design is shown in Fig. 6.2.

The box and cover as shown in Fig. 6.2 is easy to draw but not as easy to process. The box is actually a pocket; its corner radius effects processing time. A sharp 90° corner is almost impossible to make (it needs a special tool and machine unless it is made by injection) For example a small corner radius calls for a small diameter mill for processing the corners and might double, if not more, the processing time.

Note: The easiest shape to process is a round hole. Therefore if it is possible, a hole should be cut instead of a square shape, as it might reduce processing time. The volume may be computed by diameter and length. Any combination of these two parameters may serve as a satisfactory solution. A simple hole may be produced by a twist drill or insert drill. However there are limits to diameter size and the ratio of length to diameter.

The designer must be careful in assigning diameter tolerances. Drawing usually requires setting tolerances to any dimension or setting standard tolerance to all dimensions given in the drawing without tolerances. Several assign a general tolerance of ± 0.1, which unintentionally might result in an increase in processing time of about 25%.

A twist drill is not intended to process with a tolerance of ±0.1 and it calls for a second process operation to ream the hole in order to meet the tolerance specified in the drawing.

There are many factors that the designer should consider and there is no one "best design". This is demonstrated in Fig. 6.3 which shows several designs of a pulley. Each one of the designs will meet the design objectives; therefore, they all are good designs. However, from a standpoint of assembly and economics there is quite a different consideration in each design.

Design A is a traditional design. It saves assembly time but incurs extra cost of raw material and machining time.

The assembly is straightforward, all parts are assembled from one side, assembly tools and features are standard items. However, from a machining and economics point of view it results in waste of raw material, and machining time of the shaft. The raw material is 25Ø mm × 70 mm long. The shaft has to be turned to 18 Ø for a length of 24 mm.

Design B reduces the cost of raw material and machining but increases the cost of assembly.

In this design the raw material is 18Ø mm × 70 mm long instead of 25Ø mm × 70 mm long. There is no need to turn 24 mm long from 25Ø to 18Ø mm. However, there is a need to add a washer of 5 × 25Ø with a hole of 12Ø mm. The assembly is from one side only, but with an added part (the washer) for the assembly operation.

Design C reduces the assembly and machining cost but increases raw material cost.

The task washer in design B and the shoulder in design A is to reduce the surface pressure (compression stress) and wear on the moving parts. However, by choosing a more wear-resistant material, the shoulder may be reduced from 25Ø mm to 18Ø mm without affecting the lifetime of the pulley.

Design B increases the cost of assembly, as there are more parts to assemble, but the additional cost might be compensated by the reduction in raw material and machining cost.

Figure 6.4 shows how a design may reduce the cost of assembly, material and processing.

Figure 6.4 is for sure an example of over design. To prevent the disc from rotating there is no need to use two screws and nuts. There might be other methods. One of the possible designs is shown in Fig. 6.4. In this design a simple press fit hole and a spacer is used, the assembly is quite simple and straightforward.

In the absence of some formalized procedure for considering design for assembly, many factors tend to be overlooked or ignored by the individual designer who is usually not in a position to consider different methods of assembly. The DFA—design for assembly objective is to assist and guide the designer in order to remove this obstacle.

1 Introduction

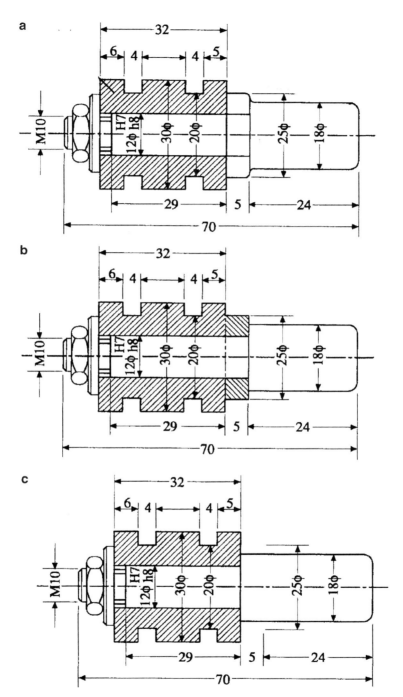

Fig. 6.3 Design of a pulley

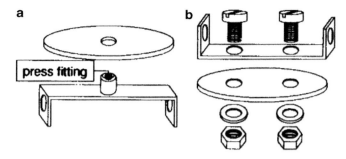

Fig. 6.4 Design of a disc assembly

1.1 Assembly Steps

Assembly is a collective term for a large number of very different technological steps. Assembly planning might use a manual—automated—robotic—conveyor while each method might have different obstacles, therefore it is difficult to guide the designer before the assembly method is selected.

Assembly has at least four stages:

- Moving
- Locating
- Laying
- Fastening

Each stage might pose different design constraints, and even conflicting demands between one stage and the other.

Design encounters many different projects and objectives. Each calls for different assembly problems. It is difficult to have an assembly-oriented theory that covers all cases.

Engineering design is an innovation process, depending heavily on the designer's imagination, talent, and experience. Many designers regard themselves as artists rather than engineers. They may regard the assembly planning as invading their territory. One may have to fight hard to gain concessions from product designers and whilst they may appear to cooperate willingly they often reserve the right to re-specify the components as the project progresses.

The following main obstacles to automating assembly are:

- Absence of assembly-oriented product design theory.
- Difficulties in handling individual parts.
- A large proportion of adaptation and adjustment work.
- Visual inspection during assembly.
- Inadequate manufacturing precision of individual parts.

The influences of assembly-oriented product design in the production process are shown in Fig. 6.5.

2 Assembly-Oriented Planning

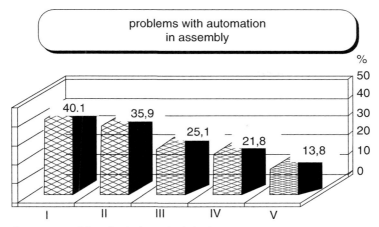

I no assembly-oriented product design
II difficult handling of components
III high portion of adaption and adjustment activites
IV Visual inspection during assembly
V low manufacturing precision of single components

Fig. 6.5 Percentage of each obstacle to automating assembly

If these points are considered more closely, it becomes evident that the poor handling aptitude of individual parts and high proportion of adaptation and adjustment work can be avoided by assembly-oriented product design. A survey also shows that if design-specific modifications were made, 60% of the parts could be assembled automatically.

2 Assembly-Oriented Planning

Assembly-oriented design is a requirement designers are faced with very often; it is intended mainly to save costs and production times as well as to assure quality in the production sector of assembly. So far little help for assembly-oriented design has been offered to designers. Such help included relatively trivial rules, such as "reduce the number of components" or "provide uniform joining directions" or comparisons between assembly-oriented and non-assembly-oriented solutions in the form of catalogues. As for the normally global rules, it must be stated that they may lead to wrong solutions in concrete applications; for example, it is quite conceivable that a multitude of simple components is more economical to assemble in a suitable automatic machine than a few complex components for which automation is not possible. It may also be assumed that when, for example, a multi-axis industrial robot is used as an assembly means, uniform directions of assembly are of minor significance. Comparisons in catalogues of solutions must be regarded with the same reservation, because the assembly means is not considered, i.e., the type of assembly. A product that has an "assembly-oriented" design in the beginning may lead to the most uneconomical overall solution in an assembly system that is not tailor-made for the product.

Thus, it is essential to state that there need not necessarily be "one assembly-oriented design" of a product, but that this depends on the assembly means to be used. The assembly methods may be classified as:

- Manual method
  - Manual assembly
  - Manual assembly with mechanical help
- Automated assembly machines
  - Special purpose indexing machine
    - Continuou transfer (indexing)
      - Rotary
      - In-line
  - Intermittent transfer
    - Indexing
      - Rotary
      - In-line
    - Free transfer
      - In-line transfer
- Robotic assembly
  - Single station robot machine
  - Multi-station robot machine
  - Robotic assembly in line with manual loading magazine
  - Robotic assembly in line with automatic feeding
  - Robot assembly cell with manual loading magazine
  - Robot assembly cell with automatic feeding
- Hybrid automatic-manual assembly system

The many assembly systems can be grouped into three classes of assembly, which are:

- Manual assembly
- Automated assembly
- Robotic assembly

2 Assembly-Oriented Planning

2.1 Manual Assembly

In manual assembly, the observation, the control of motion, and the inherent decision-making ability of the assembly operator are far superior to those of even the most sophisticated machine.

Since the assembly operator has such controlled versatility, the tools required are generally simple and less expensive than those that are necessary for any form of automated assembly. The assembly operator can identify defective parts, thus making down-time due to poor parts quality almost negligible.

Manual assembly systems are best suited to relatively low volume products that have a high product variety due to their flexibility and adaptability.

The main advantages of manual assembly are that a human assembly operator can make intuitive judgments; orientation of assembly parts requires the most difficult combinations of motion.

2.2 Automatic Assembly

There are many reasons that go toward justification of automated assembly, but the overriding one is economic productivity by cutting down assembly costs. Automated assembly is best applied to situations of high volume production of a product that has little or no product variations, involving labor intensive assembly operations. It is under these circumstances that it offers significant potential benefits in terms of assembly cost minimizations and increased productivity. The main point to note is the phrase 'special purpose'; this implies dedication, i.e. devotion to a single function. This means assembling a single product with little or no product variation. It is significant to note the large batch size and high annual production rates that are economically suited to automated assembly.

During automated assembly, individual assembly operations are normally carried out at separate workstations. The assembly, in various stages of completion, is transferred from workstation to workstation on a carrier.

Automated assembly machines are described in terms of mechanism transferring their work carriers. The types of machine available are either continuous transfer or intermittent transfer, and the motion available may be either rotary (circular path motion) or in-line (straight line motion).

2.3 Robotic Assembly

The place of robots in assembly has for several years been the subject of vigorous research but reasonably little industrial application. The trend has now changed and one of the largest growth areas of robotics applications has been in assembly operations.

By no means are all assembly operations suited to the use of robots. The versatility of robots is obviously not suited to mass dedicated production, but rather to applications where a variety of products are assembled. Where a number of varieties exist within a product family, a programmable assembly system can bring the benefits of automation to the assembly process whilst coping with the necessity of regular changeover by good software design and modular tooling.

One of the major problems associated with robotic assembly is the speed of production. This can be clearly seen by comparing the special purpose assembly machines with robotic assembly. The former is parallel assembly (each cycle of machine results in the completion of a product), whilst the later is serial assembly (each operation only results in the addition of one part of the product). Typical figures of special purpose assembly are between fifteen and sixty parts per minute, whilst those for robotic assembly are between one and four parts per minute. The figures quoted for robotic assembly appear somewhat pessimistic and are probably those for assembly of heavy parts.

The question is how a robot fits into an assembly scene. The answer lies in medium batch size assembly operation of products with several variations using low cost high speed robots. The need for speed is obvious, although this bring up the two conflicting requirements of all robots, i.e. the need to be light so movements can be rapid and consume little power, whilst at the same time the structure needs to be rigid enough to prevent deformation under load, especially at maximum extension within the robot's envelope. Similarly repeatability needs to be of a high order to maintain the positional integrity of the working assembly system.

Cheap methods of part presentation are essential in order to keep overall system cost down. In the future, the use of robots for assembly will depend on an improvement in the methods to deliver oriented parts to the robot work cell.

Low-cost assembly robots are essential in order to improve economic productivity when the capital cost is amortized over the predicted life of the robot.

In short, there are no fixed rules to determine whether or not robotic assembly should be used, but rather a careful feasibility study is required. One method of increasing confidence in a proposed system layout and selection of a robot is by employing a computer/graphical simulation package which permits simulated interaction of robot and workstation to perform assembly tasks, and optimize assembly motions and cycle times.

2.3.1 Comparing of Robots-Human and Automation

Product design will differ in case of design for human assembly; automated assembly; or robotic assembly. The following statements will assist in making the decision of selecting the assembly method:

- The more similar two jobs, the easier it is to transfer work between robots and humans. For humans, the transfer is just learning a similar task, for robots it is a matter of reprogramming. Automation assembly falls short in this case since a new machine is needed for the operation.

- Retention of facts will be less likely on humans than machines or robots. Humans forget a task after an interrupted session and their performance will be lower than before the break. The robots will retain the task in its memory and the machine, as long as functional, will be consistent in its work.
- A human possesses a set of basic skills and experience accumulated over the years and therefore requires less detailed training. Meanwhile a robot needs an extensive detailed program for every micro-activity and micro-motion. In automated assembly machines there is a need for a new machine for every different job, since the machine has no memory.
- Automation and robots have no significant individual differences, while differences, while humans have different likes, dislikes, personality, character, etc.
- Robots and automatons are designed to do a certain task, nothing is excess, while humans can apply experience and judgment. A robot needs reprogramming for any additional task.
- Robots and automatons are unaffected by social and psychological effects while a human may carry into work any anger or stress from home.
- A machine cannot do a complete task unless supervised by a human. Automatons or robots cannot be left unattended since, in case of malfunction, a human is required to attend to it and fix it or at least be aware of the breakdown.

2.4 Hybrid Automatic-Manual Assembly

The assembly process is the most complex process in modern manufacturing. While automation has made most of part manufacturing very efficient and less dependent on manual labor, the manual assembly processes still are the most used. This situation is apparently due to the complexity of products at the final assembly stage. It is also a function of the modern trend toward mass customization assembly. Even highly standardized products are offered in a suite of variants that leads to only small batches of similar products at the assembly department.

For many companies, the yearly production volume falls far below the economic volume of robot assembly. The only possibility for reaching volumes that would seem economical in assembly is to combine products of the same family where differences are moderate.

When combining several similar products into a group to form an assembly family, it is quite common to observe that the similarity in structure is large in the first stages of the assembly process. It then becomes less pronounced as the products come closer to completion. Therefore it is fairly easy to set up an automatic assembly system to cover the first part of the process, and let the final operations be done manually. A redesign of the product is often needed to obtain sufficient similarity in the basic parts of the product.

The hybrid automatic-manual assembly system offers a good opportunity to increase efficiency.

3 Design Constraints for Assembly

The importance of early consideration of product design for the assembly process is self-evident; the consequences of lack of consideration are reflected in high manufacturing costs and high labor involvement. Manufacturing engineers readily recognize the effect of design on assembly; indeed, they spend many hours resolving difficult assembly problems after the design has been approved for production. Post-approval design changes are difficult to achieve because of their high cost. It is for these reasons that a design/manufacturing interface should be established at the earliest stages of design if an optimal design is to be successfully developed.

The designer will normally concentrate first and foremost on getting the product to function within the economic limitations laid down, and then turn their focus to the assembly process. The fact that assembly is intended to be carried out by machinery will have a fundamental influence on all aspects of the design. Although the main thrust will be assembly the designer will, to varying degrees, have to bear in mind other design considerations.

3.1 Design Rules

During assembly-oriented design there are many design constraints. On the basis of a detailed analysis of all these constraints, a generic set of rules can be established that a designer should adhere to whenever possible. There are over 60 design rules that are identified in detail in the appropriate literature.

In addition a number of design guidelines of more general nature have also been developed. The following are some of them:

3.1.1 Reducing Number of Components

One of the keystone rules for design for assembly is to reduce the number of components. For each component in a product, an automatic feeding device and at least one automaton or robot would be required.

Obviously reducing the number of component parts can significantly reduce the cost of assembly automation. Parts reduction can normally achieved by combining two or more parts together or eliminating redundant parts.

Combining parts generally implies more complex components; however, it has been found that such combinations can reduce costs to such an extent that greater savings are made on parts than on reduction of assembly time.

During the part reduction exercise every single part should be carefully examined in terms of its function, and necessity in the assembly. Especially small items such as washers, screws, clips, etc. should not escape inspection, since each one included in an assembly would require feeding, orientation and locating.

3 Design Constraints for Assembly   145

3.1.2 Parts Variation

Most marketable products do not just sell in one form only; there are usually numbers of product styles, some with various additional or optional features. This variation is normally essential to accommodate for all the envisaged customer requirements. Variations of this nature are desirable from a sales point of view, but create endless problems if the product is to be assembled automatically. It is essential to know all intended variations at the outset of the design in order to prevent major problems in assembly.

If product variations are unavoidable, then as many components as possible should be made common to all product variants. These common components should, whenever possible, contain all features that are used on each product variant, even to the extent of incorporating redundancy.

Minimizing the number of product designs, and consequently part designs, will mean that fewer parts feeders will be required. When product variants are unavoidable, then the parts variants of the different products should be assembled as near the end of the assembly process as possible. This means that a common core assembly containing all common parts is assembled initially, and then the parts for the different designs are assembled last. This assumes that the product variations are only in styles or accessories; if variations in basic operation are offered, then different assembly lines will probably be necessary.

This strategy for assembly is essential for dedicated automatic assembly in large volume production, although the strategy can be relaxed for robotic assembly due to their inherent flexibility, and assembly strategy should be structured toward reducing throughput time to a minimum in order to maximize production rate.

3.1.3 Kinematics

We need to consider several design constraints that are related to the total product structure. Dedicated assembly machines usually use 'single-handed' work heads, and industrial robots are usually single-arm machines. Therefore the designer should think in terms of 'single handed operations'.

If one thinks in terms of a single-handed assembly operation, then the easiest manipulative action is to place a component down vertically onto a firm base. This is the basis of the bottom-up stack approach to assembly. Each component should be considered in a building block fashion. The starting point is a solid base that provides integral part location, transport, orientation and is strong enough to withstand the forces experienced during assembly. The base itself needs securely positioning and orienting at each assembly station position. The assembly operation is then a sandwich-like process terminating in the final component which locks the assembly together.

In the building-block approach to assembly strategy, it is important to minimize the number of assembly directions and optimize the assembly sequence.

Examination of how the parts assemble together has to be done for each proposed design. This is particularly important when parts integration is proposed and several possible configurations are being considered.

3.1.4 Placing the Component into a Product

There are two distinct areas which are important if components being assembled are to mate correctly and repeatedly. They are:

(a) Product assembly design
(b) Component design for placement

In product assembly design the product should be designed around a horizontal and a vertical datum that will provide references upon which the movements executed by any of the automatic placing or fastening mechanisms and any required calibration can be fixed. If no functional feature or features of the product can be used, then a non-functional projection or tooling reference may be necessary. Ideally, the major component of a product should act as a building nest for assembly. The rest of the components should then be placed and if necessary, fastened into position in a natural sequence without previously assembled components causing any impediment.

It is preferable that partial assembly should not be required to be moved off its jig or turned over during assembly; however, complex work-carrying jigs and a new datum position are invariably required. If turning of the assembly is unavoidable, components previously placed in the assembly but not yet fastened are in danger of moving out of position and hence temporary retainers should be specified as required. It is important to consider any visual inspection/calibration/adjustment and testing and checking required. During design, accessibility can be made available for any of these procedures and thereby make it possible to mechanize an operation which invariably in manual assembly is performed off-line.

3.2 Orientation

There is a danger that the product designer, who is used to designing products for manual assembly, will not fully appreciate how even the most unskilled human operator is quick to respond at handling parts. Although a machine can be made to simulate a human operator, the capital cost involved is usually prohibitive. In order to keep this cost to a minimum the designer should as far as possible design the required components for minimum orientation. There are in general, two classes of components which can be automatically orientated with a minimum of cost.

(a) Completely symmetrical components that are always oriented, e.g. ball bearing, cylinder, plain cube etc.

3 Design Constraints for Assembly   147

(b) If component design to (a) above is not possible, components should be as asymmetric as possible or incorporate a special feature such as a lug, notch etc. to make them asymmetric. Following these two rules will help to ensure that extremely difficult operation problems do not occur. However, the complex range of likely shapes a designer may meet is infinite and hence no complete set of rules can be given.

3.3 Fastening

The methods commonly used to fasten components together can be listed in four categories.

(a) Joining with no separate fasteners required
(b) Joining requiring one separate fastener per joint
(c) Joining requiring more than one separate fastener per joint
(d) Joining by heat, joining with no separate fasteners

3.3.1 Joining with No Separate Fasteners Required

Usually the use of pressure is required, e.g. swaging, staking, crimping, twisting and spinning. Often integral parts of the components themselves are used, i.e. integrally cast rivets, built-in clips on moldings, tongue and slot joints, etc. Because the requirement for feeders, orientation mechanisms and placing mechanisms are eliminated, this category of fastener is recommended for mechanized or automatic assembly. Pressure provides a simple, straightforward action fastening method and can be applied in numerous ways, e.g. impact squeeze, vibratory etc. using mechanical, hydraulic, pneumatic or electrical actuation. The machine designer will select which method is required. The main danger with using pressure techniques is the possibility of dislodging other unfastened components or causing damage to the product. The method of pressure application should be selected accordingly with the provision of clamps where necessary to avoid dislodging.

3.3.2 Joining Requiring One Separate Fastener per Joint

Examples of fasteners are: rivets, drive nails, screws, and self tapping screws. Adhesives can also be included in this category. Feeding and orientation mechanisms are required (or adhesive applicators) although normally relatively simple. If a separate fastener is unavoidable it is preferable for automatic assembly to use fasteners which require pressure to secure, such as rivets, drive nails etc.

A common argument used against these and for threaded fasteners is the requirement for ease of dismantling for re-work. Often this is valid especially when routine servicing is required during the life cycle of a product. Frequently, however, threaded fasteners are used purely for a re-work capability during production. Hence the majority of the products which come off the line carry a cost penalty because of a small percentage of the total production which requires re-work. Far better to design for easy removal of rivets etc. and thus only the small re-work percentage carries the cost penalty.

Adhesives are a relatively new field in assembly and their potential has not yet been fully realized. The bonds formed are strong. The main problems encountered when using adhesives are during application. The danger of blocked applicators due to premature hardening of the adhesive is very real, however the major problem to be considered is the splashing of the adhesive on other parts of the assembly.

Development in this field is continuing and the product designer should ensure that he is up to date with new developments in an area, which, when fully developed, could provide the basis of major advances in assembly techniques.

3.3.3 Joining Requiring More Than One Separate Fastener

The obvious example is the nut/bolt/washer combination. It is always preferable to feed the nut first and then drive the screw into it and in many cases it is safer to start the thread with light pressure at one station and transfer to a second station for final tightening. Combination fasteners should be avoided wherever possible in quantity mechanized assembly.

3.3.4 Joining by Heat

Welding and heat sealing; Both processes require no additional material to make the joint and as such are ideal fastening methods, but require time and constant pressure applied at the electrode during the joining process. Unlike pressure jointing, welding cannot be done as a splitting up operation and if the time cycle required is short, two or more work heads operating simultaneously at one work station may be required.

Heat sealing is used to fuse plastic components together under light pressure. As the use of plastic components continues to grow, heat sealing will be used more and more.

Soldering and brazing are high efficiency methods of joining parts. Automatic flow soldering has been with us now for many years and has dramatically increased production efficiency. Cored solder, available in various forms, e.g. discs, washers, rings, gaskets, can be fed as any other component. Equipment for feeding, cutting and shaping of cored strip on the assembly machine can reduce feeding and orientation requirements.

4 Component Design for Placement

4.1 Component Which is Nearly Identical on Both Sides

Flat components are often identical on both sides, e.g. a washer. However, occurrences of components that are not quite identical when turned over are common. In manual assembly, the difference can be highlighted by coloring or similar techniques so that the operator can see to assemble the component the correct way up.

However for automatic orientation, before conceding to the requirement for sophisticated sensing and turning equipment, the product designer should first try to design a component to be reversible. A non-functional notch ensures that correct orientation in the second plane can be easily achieved.

4.2 Headed Fasteners

Most headed fasteners such as screws, rivets, drive nails etc. will be fed between rails or down a tube. In the case of feeding between rails, special attention should be paid to the ratio of major to minor diameter of the fastener to ensure trouble free feeding.

For feeding down a tube it is essential that the major diameter does not exceed the overall length of the fastener. A head diameter to overall length of fastener ratio 1:1.25 is recommended as a minimum.

4.3 Components Design for Placement

In general, the same rules apply for automatic assembly as those for manual assembly although control must be even more stringent. Generous lead-in chamfers or radii on mating components must be consistently produced to avoid wedging or jamming.

Any shafts entering a hole should be designed to be self-centering by correct selection of shaft end shape. The choice of a point on threaded fasteners for insertion into a tapped hole which is not very accessible illustrates this point well. Square points such as achieved after thread rolling will not self center. Chamfered points will self center provided reasonable lineup is achieved. However, for automatic assembly a cone or oval ended point is recommended.

By designing a plain shaft to be relieved and allowing float in the placing head, the possibility of wedging will be minimized. This technique is used on pilot plug gauge; As far as possible, parts should be standardized, e.g. use rivets, screws, washers etc, with identical specifications thus minimizing the occurrence of wrong parts jamming feeders and tooling

5 Summary

Assembly-oriented design is an innovative and creative task, therefore it is almost impossible to formulate design procedure with logical steps. Moreover, there need not be a "one best" assembly-oriented design of a product, as it depends on the assembly means to be used (manual, flexible automated, rigidly automation). Yet reasonable pre-approved design should be established, at the earliest if an optimal design is to be successfully developed.

During assembly-oriented design there are many design constraints. On the basis of a detailed analysis of all these constraints, a generic set of rules can be established that a designer should adhere to whenever possible.

There is a great deal of research to determine the sequence of operations for planning the assembly line. Several methods, such as: simulation, computer-aided product analysis, heuristics, hierarchical control, connectivity graphs, etc. are offered. Most of these methods define the problem as: given an N-array tree representing a structure S which consists of K bodies, it is required to construct a binary tree which represents the assembling sequence of the structure.

The design is taken for granted. Post-approval design changes are difficult to achieve because of the high cost of design changes. It is for these reasons that a design/manufacturing interface should be established at the earliest if an optimal design is to be successfully developed.

Chapter 7
Management Decision Support System

Abstract A roadmap manufacturing system is basically a process that begins with only engineering input and then progresses under management direction through a variety of stages to its conclusion. Its main feature is the ability to generate routing and scheduling online without a need for the continued input of an engineer; therefore, it is able to assist management by supplying the objective information and simulations needed to make decisions of an engineering nature, such as facilities planning, expansion of manufacturing capabilities, and introduction of new manufacturing technologies.

This section does not pretend to be instruction in how to implement a roadmap system. I am not an expert in the subject. My objective is simply to show how a roadmap may assist by generating data for use in all stages of manufacturing.

1 Introduction

Management's main objective is to make profit. Therefore, the optimization criteria for management decisions must be cost, capital tie down in production, and income; we refer to these collectively as "the finance criterion". To achieve this prime objective, management relies on economic models and techniques (e.g. Total Value Analysis, ROI, etc.,) in making its decisions. Different companies will adapt different economic models. However, no matter what economic model is employed, those decisions will be made on the basis of the engineering data fed into it. Thus engineering will provide the first screening of data that will determine the manufacturing path.

Engineering, no doubt, does the best it can. However, purely engineering considerations and optimization are not always similar to ideas and policies of management. Thus the data fed to the economic model is likely to be incomplete, and uneconomical decisions may be reached even when employing advanced economical models.

Moreover, by present day methodology, not even a single engineering stage of the manufacturing process considers economics and cost as its primary objective.

All-embracing manufacturing technology utilizes a roadmap system which is basically an engineering system that assists management by supplying the information and simulations needed to make decisions of an engineering nature, such as facilities planning, expansion of manufacturing capabilities, and introduction of new manufacturing technologies.

The roadmap method establishes a network of all possible routings, while deferring to a later stage the decision of which routing to take. It incorporates technology in production management, thereby, it adds a new degree of freedom, and allows crucial decisions to be made at the right time and by qualified professionals. The decision of which routing to use can be made by a computer program in a split second, without the need to contact a process planner for changes or alternate decisions that may be needed at later stages.

The roadmap method is composed of three stages in generate routing: technology, transformation, and decision (mathematics).

- The technology stage generates the basic (theoretical) process (BP).
- The transformation stage converts the theoretical stage to a practical plan by considering specific plant resources and constructing a roadmap (RP).
- The mathematics stage fills in the details of the roadmap and generates a practical process plan routine (PP) according to any immediate requirements.

It is often called the "magic roadmap", as the same basic roadmap serves almost all manufacturing stages and activities such as: production management; scheduling; resource evaluation; resource performance measurement and establishing company level of competitiveness; etc.

Engineering stages are mostly human oriented activities and rely on planners' experience and intuition. Therefore, it is possible that sophisticated mathematical management decision models are working, unknowingly, with biased and questionable data.

Engineers are not generically economics or production management experts and therefore should not be the decision makers in these areas. There are many solutions to any engineering problem. Engineering criteria of optimization are often different than those of management. The engineer's role is simply to present feasible alternatives and let management select, by its own criteria, what they consider to be the best solution.

A roadmap is not meant to subsume all systems that supervise an industrial enterprise. The administrative functions, that is, bookkeeping, inventory, costing, personnel, purchasing, sales, and job recording, are not significantly affected by the roadmap. However, some adjustments may be, and often must be, made as a result of employing the system.

2 Plant Performance Measurement

Measuring and evaluating manufacturing performance is usually an ambiguous job, since no clear objective standards exist and there are many conflicting goals. By its nature and method of planning, a roadmap can supply objective measuring reference points. These can isolate the individual effects and be used as a universal scale for performance measurements. The reference points are:

Basic process (BP). The basic process is a fixed universal reference point. Its value is based on actual available technologies. It considers real strength and technical constraints. It assumes an imaginary resource, that is, no resource constraints are considered. Thus the basic process plan is practical from the engineering standpoint and theoretical from a specific shop standpoint. Its value does not include setup cost. Consequently, it is free from sales costs, lot sizing, grouping, and scheduling effects. It is a theoretical value that will most probably never be achieved. However, it is a fixed value, representing the state of technology. It can be used as a fixed measuring reference point.

The numerical value of the absolute theoretical optimum is a by-product of the engineering phase of process planning as discussed in Chap. 1. It is expressed in time units. For comparison of variables purposes, it is necessary to use a common denominator, which is cost. The conversion from time units to cost is accomplished by multiplying machine time by its hourly rate. An arbitrary hourly rate for the imaginary resource can be assumed. The lowest hourly rate used in the shop is recommended as the assumed value. This guarantees that the dispersion will be to only one side of the fixed reference point.

Roadmap process (RP). The roadmap optimum is a fixed specific shop floor reference point. Its value is based on the actual resources available in a specific shop. The roadmap optimum is practical from the standpoints of technology and available resources, and theoretical with regard to production and capacity planning, that is, the availability of the required resource at the required time. Moreover, it considers a theoretical optimum quantity and not the actual quantity. The theoretical optimum quantity is defined as that quantity which, when increased, will have a negligible effect on the item cost. The quantity may be set arbitrarily to high value (3,000) in order to set the transfer penalty to zero.

Practical process (PP). The practical process is the time of producing a part within a product mix. Its value is the time that production planning uses for scheduling. Its value includes: lot size; grouping, operation overlapping; and balance of the work center load.

Planned performance (PF). Production planning and scheduling transforms plant orders into a processing plan. This plan is transferred to the shop floor for execution. The planner's schedule for available resources, however, is not necessarily final. There is competition for resources that the planner has to take into account. A feature of the system is the ability of the planner to offer various solutions to such problems;

an example might be the possibility of management selecting the proper resources for a given order but not in appropriate quantities.

Actual planned performance (AP). The actual total expenses of the department or work center and the deviation from the plan in terms of items and quantities constitute actual performance. The job recording system should supply data about completion of planned jobs and about accumulated cost in producing the product mix.

Ratios of the above reference points indicate the performance level of separate functions in the company.

2.1 Reasource Suitability to Products

The absolute theoretical optimum is a value that probably cannot be achieved. It is based on an imaginary resource that probably does not exist in any specific plant. The ratio R_M is given by the equation

$$R_M = \left(\sum_{i=1}^{i=I} \frac{BP_i}{RP_i}\right) \bigg/ I$$

where: R_M = The suitability of resources to product mix
   BP = basic process
   RP = roadmap process
     I = Number of participate products (It is recommended to use major products)

The lower the value of R_M the better the performance obtained. It indicates that more suitable resources are available for production of the given product mix.

In a way, it is an absolute ratio, as it is independent of the product mix.

It can be used to compare the level of management performance at different plants. It evaluates management decisions concerning resource planning.

2.1.1 Suitability of Resources to Company Products

For this purpose, equation parameters should be as follows:

BP = List the major products you wish to evaluate. Call their product structure and run the product items from top to bottom, calling a roadmap to compute the basic time (BP) of each item on the list. The sum of all items on the list is the value to use in the equation.

RP = List the major products you wish to evaluate. Call their product structure and run the product items from top to bottom, calling a roadmap to compute the process time (RP) of each item on the list. The sum of all items on the list is the value to use in the equation.

2 Plant Performance Measurement

I = While running BP or RP as before, count the number of items used. This number is I; as the evaluation results may be performed in a few seconds it is recommended to list all products and items.

The manager might wish to evaluate the resources not suitability but rather on processing cost, in which case the call for a roadmap should so specify. The results will serve the purpose.

Evaluating the Results

It is recommended that, instead of only the evaluated value on a total product, additional data be compiled indicating the effect of each product or even of each item of the product.

Such additional data could be easily accessed if the results of each step as described previously were recorded on a spread sheet.

The spread sheet is capable of presenting results in a graph mode format suited to a user request.

Once getting evaluations of all products, details may be drawn to show the effect of each product separately. Once each product is evaluated, the effect of each item of the product may be shown and the effect of each operation on the results.

Such data may clearly point to the profitability rate of each product and confirm the fitness of each resource in the plant.

Sorting the products by descending cost order may motivate marketing to direct their sales force to solicit orders for the products on the top of the list.

2.2 Production Planning Performance Level

By definition, the basic process (BP) results in the minimum item processing time and cost. Its value is specific to each plant and probably cannot be met. The product mix, that is, the customer orders and their due dates, affects performance and increases cost. The ratio Rs is given by the equation

$$R_S = \left(\sum_{i=1}^{i=PM} \frac{RP_i}{PF_i} \right) \bigg/ PM$$

where: Rs = Production planning performance ratio
       PF = Practical process to produce a product mix
       RP = Actual production to produce a product mix
       PM = Product mix

The higher the ratio, the better the performance obtained. It indicates that the manufactured quantities are closer to optimum, due dates were set in such a manner that less competition for resources occurs, product mix is such that it allows separate items to be grouped in a single manufacturing batch, and there are sufficient orders and flexible due dates to allow the work centers to be loaded to their full available capacity.

These factors can partially be controlled. It is up to sales to accept a sufficient number of orders, to promise realistic due dates, and to direct their sales effort to orders that complement each other in terms of load.

The master production plan can be used as a guide to sales effort and establishing realistic due dates. Some future periods are left under loaded in the master production plan. When needed, a load can be pulled forward within the allowable range and can supply sufficient work for the frozen periods. This is not, however, a way to induce sales people to promise unrealistic early deliveries of the unavoidable rush orders.

The Rs ratio can also be used to study different loading methods employed by production planning. A dispatching rules study might use this ratio as an objective measuring scale of the performance of each rule.

2.3 Shop Floor Performance Level

The ratio R_F indicates the performance level of the shop. It can be used to evaluate foremen. The ratio R_F is given by the equation

$$R_F = \left(\sum_{i=1}^{i=PM} \frac{PF_i}{AP_i} \right) \bigg/ PM$$

where: R_F = Shop floor performance level ratio
PF = Practical process cost
AP = Basic process cost
PM = Job list

The performance level is an indicator of how the foreman solves disruptions. The shop is responsible for the completion of all planned jobs (PF) within the period (AP). The ratio indicates whether this task was accomplished.

3 Resource Planning

Management has to make decisions concerning investments in equipment. Such decisions establish plant level performance and thus the ability to compete on the market. The need to make such decisions frequently arises because of:

- **Replacement of old equipment.** The life of a resource is estimated at 10–20 years. This means that 5–10% of the equipment has to be replaced every year.

- **Added manufacturing power.** When the master production plan shows a continuous overload situation, management has to decide on expansion or to turn down orders. If it decides to expand, new equipment must be purchased.
- **Disposal of unsuitable, inefficient equipment.** When a resource is continuously underloaded, or not selected as the first alternative for any product, management has to decide whether to dispose of it. The master production plan and roadmap are used to draw management's attention to such resources.
- **New products.** New products might require resource capabilities unavailable in existing equipment.
- **Technological changes.** New design technologies and new materials might call for new types of equipment. Management can use the design features of a roadmap, as described in Chap. 1, to evaluate such new manufacturing technologies as precision casting or molding instead of machining and bonding instead of welding. When such technologies are adopted, an automatic change is made in all company product designs and drawings, the technology is added to the engineering stage of the roadmap, and new equipment is needed.
- **A new generation of resources is introduced.** Management has to keep track of technological developments. The new generation of resources should be evaluated from a technical and commercial standpoint.

The competitive environment of an industrial enterprise must continually improve its manufacturing techniques and manufacturing resources. Research shows that about 90% of the incurred cost of manufacturing is established in the manufacturing planning stages. Therefore, the selection of the manufacturing resources and manufacturing technique is a crucial one as processing efficiency establishes plant level of performance and thus the ability to compete in the market.

In selection of new manufacturing resources considerable judgment must be exercised to assure sound decisions. The manufacturing resources must have the capacity and the other technical operating characteristics that will enable it to perform the required job plans. It must also be economically justifiable on the basis of savings in the various applicable elements of cost.

New resources usually possess more capabilities than old ones. If optimum processes are to be used, all company routings should be examined and new process plans prepared. Merely replacing a resource number in the routing file will result in inefficient manufacturing methods. However, it is impractical, by using today's techniques, to prepare a new set of process plans whenever a resource is added to the plant. It is a huge job and seldom done in general practice. Processes that might benefit remain unchanged. Thus the data fed to the economic model are incomplete.

The roadmap system does not suffer from such a degradation of manufacturing efficiency. Routings are not stored in files. The process plan is recomputed whenever needed by the presently available resource list.

The trend in resource development is toward computerized high power machines, and toward industrial robots, AGV, automatic welding, and machining centers, FMS, to name a few. The new resources are better qualified and more efficient, but their price is accordingly high. There is no doubt that employing such modern resources may save setup time, increase uptime and quality, reduce material handling, and simplify production planning.

However, it is questionable if they reduce production cost in all cases. In many cases a 35 KW machine center with five degrees of freedom, that costs about $150,000, is employed in drilling a series of ¼" holes. Such an operation may be carried out more efficiently by a $1,000 drill press.

In a metal cutting process a rough cut usually precedes the finish cut. A rough cut does not require accuracy and may be produced by an old inaccurate resource, which probably was fully depreciated. Employing modern resources, for all operations, no doubt will reduce manufacturing time and result in ease of managing. However, modern resources not always will result in the minimum production cost. Therefore, reevaluation of process planning of all products should be made and management supplied with alternative data. It is up to management to make an operating decision, but decisions must be based on sound data. The need to make decisions concerning the purchase of new equipment may follow production needs.

Management relies on economic models and techniques (e.g., total value analysis) in making its decisions. No matter what economic model is employed, the first decisions are restricted to the engineering data fed into it. Thus, as pointed out above, engineering provides the first screening of data, but does not reveal its reasons for decisions or possible alternatives and may have different criteria from those of management, thus risking an incomplete model leading to uneconomical decisions Engineering, no doubt is doing the best they can. However, the engineer is a process planning expert, but not expert in economy, engineering consideration and optimization criteria are not always similar to those of management. Thus the data fed to the economic model are incomplete, and uneconomical decisions might be reached even by employing advanced economical models.

To improve decision making, process technology should be introduced into the economic model. The engineering task will be only to supply data and not to make decisions. This task is accomplished by the roadmap method. The following section will demonstrate how it is done.

For decision purposes, a roadmap can be formulated in one of two options: resource recommendation or resource evaluation.

3.1 Resource Recommendation Module

A roadmap is divided into two major parts. The engineering stage is based on a comprehensive knowledge of the capabilities and limitations of available technologies. It begins with computation of a process to be carried out on an imaginary resource. The output of this stage is an operations list; the characteristics of the imaginary resources are precisely specified. The types of machining operations, such as turning, drilling, reaming, grinding, and milling, are specified. The power, speed,

3 Resource Planning

and feed needed are specified in the appropriate columns. The special attachments required are specified by codes. The physical size of a resource is dictated by the part dimensions.

The operations list can thus be regarded as a resource specification sheet. These specifications disregard economic considerations, that is, resource cost and its load forecast. An economic model should be used to transform the above ideal resource data into a practical solution. Such a model should include resource cost, product quantities, and facilities loading. Any plant can use the above data in its own economic model.

The roadmap method is presented in Chap. 1. It constructs a matrix in which the basic operations required to produce a part are listed in the first column, while the available resources are listed in the following columns.

The second stage is transformation of the basic operation processing time to the processing time on any one of the listed available resources.

The third stage is a mathematical program that generates routing.

The same method may be used for resource recommendation. When the need to purchase new resources arises, the process planner is asked to recommend the required resources. A list of alternate resources is assembled, usually based on catalogs, information provided by visiting salesmen, specifications of previously utilized resources, or knowledgeable random searches. The process planner has to evaluate the process plan for each one of the candidate resources, and transfer recommendations to management for economics based decisions. Process planning is a laborious task, and it is expensive to evaluate many alternate resources. Therefore, the process planner proposes a limited number of alternatives (if at all) and leaves it to the economist to decide which one of them to select. Hence, the "best" alternative might not be considered, and a biased decision might be reached.

The roadmap represents all possible processing methods. It includes an almost infinite number of process plans without pointing to the "best" process. This is done because the term "best" is an uncertain one and depends on the criteria employed in optimization. For the process planner it might mean the process that will result in a minimum process time, while for the economist it might mean minimum cost of the component, while for management it might mean a maximum profit process plan.

The basic Process (BP) is theoretical from a specific shop's viewpoint, but must be concrete from a technology standpoint. It cannot violate any physical or technological rule. In this sense the BP indicates all the characteristics and features that are most desirable to have in the resource.

Therefore the BP process is generated on an "imaginary resource". The term "imaginary resource" might be frightening. It is a resource with unlimited power, infinite speed etc. However, one does not have to be alarmed. A varying number of operations are required to produce a part. There are roughing operations that require heavy forces and limited accuracy, while a finish operation usually requires light forces but a significant accuracy. The process considers many real constraints such as part specifications, part shape and strength, fixture etc. Therefore most operations will require commercially available resources, and only a few operations might require special resources.

Each operation specifies the power, moment, forces, speed, revolutions per minute, feed rate, size of part, the accuracy required by the operation etc. These data actually point to the "best" characteristics that a resource should possess in order for the particular operation to be performed in the most economical way.

Therefore, the needs of the individual BP operations, will be used as specifications for RFQ—Request For Quotation—that will be distributed to resource building companies and suppliers.

The quotations received, in response to the RFQ stage, following a quick review, will be entered as the heading of the matrix, meaning that a process plan must be generated as if these resources are available at the specific shop. The BP process operations and priority codes are entered into a matrix.

At this stage the characteristics of each individual resource is known by the quotation received. The technical capabilities of each resource is considered (such as: power, speed, moment etc.) in transforming the BP time to the time to process each operation on each resource and in constructing the matrix.

The economic model may vary from one plant to another. However, the basic data that goes into the model are similar. The required general data might include: resource cost, finance cost, installation cost, maintenance cost, energy consumption cost, labor cost, life cycle, etc. These data are available from the quotation supplied by the resource manufacturer and the plant's accumulated economic experience. The required technical data includes the machining time per part; the cost of machining a part; anticipated rejects percent; resource utilization per part, product or product mix. This data can be furnished by the matrix.

3.1.1 Example of a Roadmap

The following example demonstrates the power of employing the matrix as a smart resource planning. Assuming the resources have to be purchased to produce part "CROSS", as shown in Fig. 7.1, the Basic Process (BP) is shown in Fig. 7.2.

The BP shown in Fig. 7.2, is the basis for a distributed RPQ—request for quotation. As a result of the RPQ, six resources were selected as candidates for purchasing. Simplified specifications of these resources are shown in Table 7.1 and are the basis for constructing the matrix, as shown in Table 7.2.

Normally the matrix solution recommends the "best" process plan. However, in this application, the target is to evaluate cost—performance of the alternative resources. The role of the matrix is to supply objective data to management, who will make the decision. To accomplish this task a computer program was created to generate many alternate processes, using different resource combinations. The purpose of generating alternatives is to prepare data that reflects machining time and cost, as a function of the investment in purchasing a new resource. The solution generated 22 alternate processes employing different resources and resource combinations. These 22 alternate processes are shown in Table 7.3.

3 Resource Planning

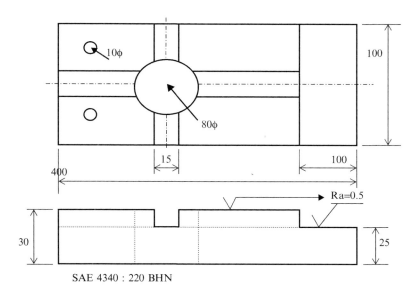

SAE 4340 : 220 BHN

Fig. 7.1 Sample part "CROSS"

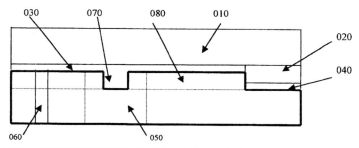

No.	Operation	Priority	Tool Dia	Leng mm	Dept mm	Feed *	Speed m/min	Power KW	Time Min
10	Rough milling	0	125	428	4.8	1825	100	32.6	0.30
20	Rough milling	010	125	293	5.0	1324	100	12.3	0.22
30	Finish Milling	020	125	528	0.2	477	156	0.72	1.11
40	Finish Milling	020	125	228	0.2	477	156	0.22	0.60
50	Center Hole	010	80	30	-	0.35	139	39.4	0.16
60	2 Hole 10φ	010	10	30	-	0.18	15	0.57	0.84
70	Short slot	050	15	80	5	178	21	1.25	0.52
80	Long Slot	050	15	240	5	178	21	1.25	1.36

* Feed rate for milling is mm/min, for drilling in mm/rev   process was generated by CAPP

Fig. 7.2 Basic process plan of part "CROSS"

Table 7.1 Resources for "CROSS"

RPQ	Machine description	Power KW	Speed RPM	Handle time min.	rate $/min	Machine cost $
1	Machining Center	35	1,500	0.10	4	150,000
2	Large CNC Milling	35	1,200	0.15	3	112,500
3	Manual machine	15	1,500	0.66	1.4	26,250
4	Small drill press	1	1,200	0.66	1	1,875
5	Old milling machine	15	2,400	1.0	1	7,500
6	Small CNC machine	10	3,000	0.25	2	37,500

Table 7.2 Roadmap operation—time matrix

Operation	BP	Priority	RPQ #1	RPQ #2	RPQ #3	RPQ #4	RPQ #5	RPQ #6
010	0.30	00	0.40	0.45	1.35	99	1.69	1.24
020	0.22	010	0.32	0.37	0.88	99	1.22	0.54
030	1.11	020	1.21	1.26	1.77	99	99	1.36
040	0.60	020	0.70	0.75	1.26	99	99	0.85
050	0.16	010	0.29	0.34	1.14	99	2.58	4.70
060	0.84	010	0.94	0.99	1.50	1.5	1.84	1.09
070	0.52	050	0.62	0.67	1.18	99	1.52	1.62
080	1.36	050	1.46	1.51	2.02	99	2.36	1.09
Total	5.11		5.94	6.34	11.10	–	–	12.49

3.1.2 Example of Decision Making

The decision of which resources to purchase is up to management. The engineering role is only to present alternatives. Simple examples will demonstrate the power of the matrix method in preparing alternatives.

Table 7.4 is constructed as an example of data for economic consideration in the decision process. The Alternate column contains an identifying number of the alternative, for which the participating resources are given in Table 7.1.

Column A contains the total cost of machining part "CROSS".

Column B contains the total time of machining part "CROSS".

Column C contains the cost of ALL resources that participate in the alternative process.

Column D contains the time to machine part CROSS on the main resource, i.e. the resource with the maximum processing time (it is taken from Table 7.3). Since resources may work in parallel, the machining rate (number of items produce in a specific period of time) should consider this time and not the total time.

Examining these columns indicates that:

Alternative 1 should be selected if the criterion of decision is maximum production.
 Thus proposal RPQ#1 should be purchased.
Alternative 3 should be selected if the criterion of decision is minimum investment.
 Thus proposal RPQ#3 should be purchased.

3 Resource Planning

Table 7.3 Alternate processes by using different resource combinations

N	Totl cost	Total time	Max. time	Resour costs $	RPQ #1	RPQ #2	RPQ #3	RPQ #4	RPQ #5	RPQ #6
1	23.76	**5.94**	5.94	150,000	5.94					
2	19.02	6.34	6.34	112,500		6.34				
3	15.54	11.10	11.10	**26,250**			11.1			
4	24.98	12.49	12.49	37,500						12.5
5	19.80	5.87	4.48	187,500	4.48					1.09
6	17.70	8.94	5.72	157,500	2.92				5.72	
7	22.80	6.80	5.00	151,875	5.00			1.50		
8	19.57	6.59	5.89	262,500	0.40	5.89				
9	16.97	6.22	4.83	150,000		4.83				1.09
10	15.53	9.19	5.72	120,000		3.17			5.72	
11	18.15	7.45	5.35	114,375		5.35		1.50		
12	15.18	9.19	7.73	138,750		1.16	7.73			
13	14.98	12.42	6.40	33,750			6.40		5.72	
14	15.19	10.47	9.08	63,750			9.08			1.09
15	24.90	13.20	11.40	39,375				1.50		11.4
16	16.09	13.04	9.99	45,000					9.99	2.75
17	14.94	12.38	6.40	35,625			6.4	1.50	3.88	
18	17.66	8.90	3.88	159,375	2.92			1.5	3.88	
19	14.90	9.49	6.23	140,625		1.16	6.23	1.5		
20	14.34	10.51	5.72	146,250		1.16	3.03		5.72	
21	14.83	9.36	6.85	165,250		0.82	6.85			1.09
22	**14.30**	10.47	3.88	148,125		1.16	3.03	1.5	3.88	

Table 7.4 Example of economic consideration

Alt.	A	B	C	D	E	F	G	H	I	K
1	23.76	**5.94**	150,000	5.94	24,038	142,788	1.19	1.25	1	7.50
2	19.02	6.34	112,500	6.34	10,246	64,959	0.54	1.33	1	5.63
3	15.54	11.10	**26,250**	11.10	1,815	20,150	**0.17**	2.33	2	2.63
4	24.98	12.49	37,500	12.49	7,470	93,302	0.78	2.62	2	1.88
5	19.80	5.87	187,500	4.48	18,382	107,904	0.90	1.23	1	9.38
6	17.70	8.94	157,500	5.72	12,805	114,476	0.95	1.88	2	15.75
7	22.80	6.80	151,875	5.00	21,094	143,438	1.20	1.43	1	7.59
8	19.57	6.59	262,500	5.89	**25,168**	165,856	**1.38**	1.38	1	13.13
9	16.97	6.22	150,000	4.83	11,512	71,604	0.60	1.30	1	7.50
10	15.53	9.19	120,000	5.72	8,293	76,213	0.64	1.93	2	12.00
11	18.15	7.45	114,375	5.35	9,652	71,907	0.60	1.56	1	5.72
12	15.18	9.19	138,750	7.73	9,362	86,040	0.72	1.93	2	13.88
13	14.98	12.42	33,750	6.40	2,247	27,908	**0.23**	2.60	2	3.38
14	15.19	10.47	63,750	9.08	4,305	45,068	0.38	2.20	2	6.38
15	24.90	13.20	39,375	11.40	7,721	101,912	0.85	2.77	2	3.94
16	16.09	13.04	45,000	9.99	3,235	42,185	0.35	2.73	2	4.50
17	14.94	12.38	35,625	6.40	2,366	29,285	**0.24**	2.60	2	3.56
18	17.66	8.90	159,375	3.88	12,915	114,946	0.96	1.87	2	15.94
19	14.90	9.49	140,625	6.23	9,313	88,380	0.74	1.99	2	14.06
20	**14.34**	10.51	146,250	5.72	9,339	98,154	0.82	2.20	2	14.63
21	14.83	9.36	165,250	6.85	10,893	101,960	0.85	1.96	2	16.53
22	**14.30**	10.47	148,125	3.88	9,435	98,781	**0.82**	2.20	2	14.81

Alternative 22 should be selected if the criterion of decision is minimum machining cost. Thus proposal RPQ#2; 3; 4; 5 should be purchased.

Notice that alternative 22 is mathematically selected, however alternative 20 is practically the same as alternative 22.

Making a decision with only the above criterion does not consider the quantity demands and the lead time to meet delivery dates.

Supplemental information is given in the following columns.

Column E contains the number of items that should be produced with each alternative, in order to arrive at ROI (Return on Investment), in case that the selling price of item CROSS is $30. It is computed by dividing the investment cost (column C), by the profit on each sale of a single item, i.e. $30 minus the total machining cost (column A). The $30 was taken arbitrarily, and a check for any other selling price may easily be computed.

Column F contains the ROI time in minutes, i.e. the time that it takes to resource the quantity that results in ROI (column E). It is computed by multiplying this quantity by the maximum time to resource item CROSS on the main resource (column D).

Column G contains the number of years that the resources must work in order to attain ROI. It converts the time in minutes (column F) to years. Assuming that there are 250 working days a year, 8 h per working day, 60 min per hour = 120,000 min per year.

Examining these columns indicates that:

Alternative 13 should be selected if the criterion of decision is fastest ROI. Thus proposal RPQ#3 & #5 should be purchased.

The quantity consideration indicates that:

Alternative 3 should be selected for low quantities of up to let us say 5,000 items. Thus proposal RPQ#3 should be purchased.
Alternative 13 should be selected for quantities of up to let us say 6,000 items. Thus proposal RPQ#3 & #5 should be purchased.
Alternative 17 should be selected for quantities of up to let us say 7,000 items. Thus proposal RPQ#3; #4; & #5 should be purchased.
Alternative 22 should be selected for higher quantities Thus proposal RPQ#2; #3; #4; & #5 should be purchased.

The quantities for the ROI are varied for each alternate process plan. The quantities are specified in column E. Column H adjusts the ROI to a common denominator.

Column H contains an adjustment of the ROI to the alternative that results in the maximum number of items to ROI. In this case it is alternative 8 with 25,168 items. The adjustment is computed as a ratio of quantity multiplied by the years to ROI of each alternative. For example, to produce 25,168 items with alternative 8, the ROI will remain 1.38 years.

However, for alternative 18 with individual ROI of 0.96 it has to produce 16,915 items. To adjust it to the quantity of 25,168 the ROI is adjusted to $(25,168/16,915) \times 0.96 = 1.87$ years. By this adjustment there is a common denominator for comparing ROI.

The value in this column indicates that alternate 22 should be selected.

The quantity considered in the above columns was based on purely mathematical considerations and not on commercial data. For commercial consideration the required quantity is supplied to the system for evaluation (also the selling price of the item, for which data may be acquired by market research). In the example it is assumed that the selling price, as before, is $30 and the required quantity is 20,000 items in one year.

Column I computes the number of resources of each alternative that must be purchased in order to meet the required quantity. The number is rounded to an integer value. It is computed by multiplying the quantity, i.e. 20,000 by the machining time as appears in column D and divided by 120,000 min per year.

Another indication for decision making might be the investment per item. Such value is given in the next column.

Column K computes the required investment per item, in case that the required quantity is 20,000. The investment cost per alternative is multiplied by the number of resources required to produce 20,000 items per year, as given in column I, and divided by the quantity of 20,000 units.

For this criterion of decision, alternative 13 or alternative 17 should be selected.

The above example does not intend to specify how management should make its decisions. Its purpose is merely to demonstrate that, when the criteria are determined by management, the matrix method is capable of supplying objective data, without the need to turn to the process planner for data for each criterion.

The above example demonstrated how to build a decision support table (Table 7.4) for a single part. In practice, multi-parts should be considered. However, the concept remains the same.

The process planner is asked to generate a process plan for each part separately, using imaginary resources. This process plan is regarded as the BP (basic process). The requirements for the BP operation are used to specify the RFQ. The received proposals are used to construct a matrix for each item separately, however the resources in all matrices are the same. Alternate process plans are generated by employing the resources of the matrices in different combinations. The alternative resources are the same for all parts to be evaluated. The alternatives for each part are organized in tables similar to Table 7.2.

For multi-parts a table of alternatives in the form of Table 7.4 is constructed in which, for each alternative line, data of all parts are entered. Considering the quantity and selling price of each part will be used for computation of ROI, investment per part in a similar method as that of a single part.

3.1.3 Summary

Summing up the alternatives, as detailed in this section, reveals that the decision of which manufacturing resources to purchase in order to produce a specific part or a product mix is not a straightforward decision. Selecting resources on the basis of a criterion of minimum cost or maximum production options is not the best decision in all cases.

The possible decisions for the above example are as follows:

RFQ #1 should be selected if the criterion of decision is maximum production.
RFQ #3 should be selected if the criterion of decision is minimum investment.
RFQ #2, #3, #4, #5 should be selected if the criterion of decision is minimum machining cost.
RFQ #3, #5 should be selected for medium quantity demand.
RFQ #3, #4, #5 should be selected for high quantity demand.

4 Maximum Profit Criterion of Process Planning Optimization

The routing in the roadmap method is considered a variable, to be selected by solving the roadmap at the time of need. However, for some application a starting routing is needed. The expectable criteria of optimization are maximum production

4 Maximum Profit Criterion of Process Planning Optimization

or minimum cost. Examining columns A and B of Table 7.4 shows that these two criteria result in a completely different routine. The question of which one to use is a management decision and not an engineering one. It usually depends on the number of orders, plant load, and seasonal products. It makes sense that in a normal period, when there is ample time to meet a delivery date, the minimum criterion of optimization will be used. For rush orders, such as a product that is specific to a certain holiday (after which the product is usually unsalable), if the delivery date is in danger, the maximum production criterion of optimization will be preferred. Another consideration is to use, in the scheduling and capacity planning system, a routing that may solve bottleneck problems such as overloaded or under loaded resources.

Some companies have proposed a ***maximum profit*** criterion of optimization as a starting routing. However, this criterion is mathematically undefined and depends on the selling price of the product, which is not an engineering parameter, therefore it is seldom used.

The roadmap method proposes and enables routings that result in a maximum profit criterion of optimization. As shown in Table 7.4, the generated process plan indicates the processing time and cost of each alternative. These two values are used to select the routing; if time is important, then use the maximum production criterion of optimization routing; if the processing cost is important, then use the minimum cost criterion of optimization. Clearly, in the use of a maximum profit criterion of optimization, both these values are considered.

The profit of a single item is the difference between the net selling price (Sale) and the processing cost (Cost). Single Item Profit (SIP) is therefore computed by:

$$\text{SIP} = \text{Sale} - \text{Cost} \tag{7.1}$$

where cost is the value retrieved from Table 7.4.

The total profit in a period is the SIP—Single Item Profit—multiplied by the number of items produced in the period (Quantity). The quantity that will be produced during the period is the number of minutes during the period (PM) divided by processing time in minutes to produce an item (RT) thus:

$$\text{Quantity} = \text{PM}/\text{RT} \tag{7.2}$$

where processing time (RT) is the solution of the time-matrix.

The Total Profit in a Period (TPP) is then

$$\text{TPP} = \text{Quantity} * \text{SIP}. \tag{7.3}$$

Equation 7.3 may be written as

$$\text{TPP} = (\text{PM}/\text{RT}) * \text{SIP} = \text{PM} * \{(\text{Sales} - \text{Cost})/\text{RT}\}. \tag{7.4}$$

Equation 7.4 indicates that the process alternative that results in total profit per period (TPP) is actually independent of the length of the period, as it is a linear function of the length of the period. The engineering variables in (7.4) are Cost and Time (RT) and the management variable is the Sales price. Therefore, in computing the maximum profit process plan we may ignore the length of the period (PM) and use instead a fixed value of profit at a unit period which is a Relative Total Period Profit (RTPP) value.

$$\text{RTPP} = (\text{Sales} - \text{Cost})/\text{RT}. \tag{7.5}$$

To demonstrate the effect of the process plan alternative on the profit let use again the matrix of part "CROSS" as shown in Table 7.3.

For an economic lot size of 125 units per batch, and a setup cost of 30, retrieved values from alternative 22 are: Minimum cost of 14.30 and processing time of 10.77 min.

In a period of 8 h, 480 min quantity of items produced will be 480/10.77 = 44.57 items. If the selling price of each item is 30, then the daily profit will be: 44.57 * (30 − 14.30) = **699.75**.

Using the maximum production process (alternative 1) with the above conditions will result in an item cost of 23.76 and processing time of 5.94 min.

The number of parts that can be produced per day is 480//5.94 = 80.8 and the daily profit is 80.8 * (30 − 23.76) = **504.19**.

Using an alternative process that is not maximum production and not minimum cost but an arbitrary data (alternative 11) with cost of 18.15 and processing time of 7.45 will result in 480/7.45 = 64.43 products per day and a daily profit of 64.43 * (30 − 18.15) = **763.49**.

The problem is to find an alternate process plan that will result in maximum profit.

5 Determining a Process for Maximum Profit

As we have noted before, engineers are generally not qualified to set product selling prices, and management is generally not qualified to generate a process plan. Engineering should supply data to management in order to make decisions based on real facts and not on assumptions. Following the roadmap system concept in determining the process that will result in maximum profit is therefore divided into three stages:

The first stage is to generate data.
The second stage will be market research.
The third stage incorporates the first two stages to determine a maximum profit process plan.

5 Determining a Process for Maximum Profit

Table 7.5 Relative total period profit (RTPP)

Alt	Cost	Time	Sale 20	25	Price 30	35	40	45
1	23.76	5.94	−0.633	0.209	1.051	1.892	2.734	3.576
2	19.02	6.34	0.155	0.943	1.732	2.521	3.309	4.098
3	15.54	11.1	0.402	0.852	1.303	1.753	2.204	2.654
4	24.98	12.49	−0.399	0.002	0.402	0.802	1.203	1.603
5	19.8	5.84	0.034	0.890	1.747	2.603	3.459	4.315
6	17.7	8.94	0.257	0.817	1.376	1.935	2.494	3.054
7	22.8	7.1	−0.394	0.310	1.014	1.718	2.423	3.127
8	19.57	6.59	0.065	0.824	1.583	2.341	3.100	3.859
9	16.97	6.22	0.487	**1.291**	**2.095**	**2.899**	**3.703**	**4.506**
10	15.53	9.19	0.486	1.030	1.575	2.119	2.663	3.207
11	18.15	7.45	0.248	0.919	1.591	2.262	2.933	3.604
12	15.18	9.99	0.482	0.983	1.483	1.984	2.484	2.985
13	14.98	12.42	0.404	0.807	1.209	1.612	2.014	2.417
14	15.19	10.47	0.459	0.937	1.415	1.892	2.370	2.847
15	24.9	13.5	−0.363	0.007	0.378	0.748	1.119	1.489
16	16.09	13.34	0.293	0.668	1.043	1.418	1.792	2.167
17	14.94	12.98	0.390	0.775	1.160	1.545	1.931	2.316
18	17.66	8.9	0.263	0.825	1.387	1.948	2.510	3.072
19	14.9	10.09	0.505	1.001	1.497	1.992	2.488	2.983
20	14.34	10.81	0.524	0.986	1.449	1.911	2.374	2.836
21	14.83	9.36	**0.552**	1.087	1.621	2.155	2.689	3.223
22	14.3	10.77	0.529	0.994	1.458	1.922	2.386	2.851
MAX	RTPP		**0.552**	**1.291**	**2.095**	**2.899**	**3.703**	**4.506**

5.1 First Stage

The equation for computing the Relative Total Period Profit looks very simple; however, as shown in Table 7.3, each alternative results in a different Cost and processing Time (RT). Similarly the selling price may be regarded as a variable to be determined by management.

To establish the optimum alternative selling price and the alternative that will result in maximum profit, the following algorithm is proposed.

Use (7.5) to compute RTPP for a range of selling prices from 20 to 45 unit cost. The results are shown in Table 7.5.

The maximum value in each column (selling price) points to the process alternative that will result in maximum profit. (It is marked by bold figures.)

It is quite surprising that above a certain selling price the optimum process plan remains constant, i.e. alternative 9. This can be explained by examining (7.5) which shows two elements:

$$\text{RTPP} = (\text{Sales} - \text{Cost})/\text{RT} = X - Y,$$

where $X = \text{Sales}/\text{RT}$ and $Y = \text{Cost}/\text{RT}$.

Table 7.6 Market research results

Sale price	20	25	30	35	40	45
Quantity	10,000	9,200	8,500	7,300	5,000	4,000
Relative quantity index	1	0.920	0.850	0.730	0.500	0.400

To get an RTPP maximum value, X must be as high as possible, which means a high selling price and minimum process time, while Y must be as small as possible, which means a minimum process cost and maximum processing time. On one hand (X) the processing time must be maximum while on the other (Y) minimum. Therefore a compromise must be found.

Our intention is to work with relative indexes, therefore a relative maximum RTPP value was added in the lower row, assuming that the maximum value of the selling price of 20 is 100%. The relative index is computed by dividing the appropriate RTPP by the RTPP of the selling price of 20; its value is shown in the last row of Table 7.5.

5.2 Second Stage

In determining the process that will result in maximum profit, the selling price must be known.

The selling price affects the sale quantity as well. Such data may be obtained by market research where the problem is: If the product selling price is Z, how many units per period (let us say a week) can be sold at each one of the indicated product selling prices. Let us assume that market research obtained results as shown in Table 7.6.

The table shows that as the price goes up the sales quantity goes down, which is reasonable. Our intention is to work with relative values, therefore a relative quantity index is added, assuming that the quantity at the selling price of 20 is 100%. The relative value is indicated in the third row of Table 7.6.

5.3 Third Stage

The selling price that will result in maximum profit per period is a function of the quantity required per period, and the routing that will result in maximum profit per selling price. The two parameters act in opposite directions.

Therefore, there is an *optimum selling price* that will result in maximum profit per period.

As these two parameters have different dimensions they cannot be used for computation of the optimum. Therefore, both are transformed into dimensionless values which will be called an index.

5 Determining a Process for Maximum Profit

Table 7.7 Integration of engineering and marketing data

Sale price	20	25	30	35	40	45
Alternate process no.	21	9	9	9	9	9
RTPP max. value	0.552	1.291	2.095	2.899	3.703	4.506
Relative alt. index	1.0	2.339	3.793	5.251	6.708	8.163
Relative quan. index	1.00	0.920	0.850	0.730	0.500	0.400
Result index	1.00	2.152	3.224	**3.833**	3.354	3.265

Table 7.8 Verifying algorithm results

Sale price	20	25	30	35	40	45
Process alt. no.	21	9	9	9	9	9
Processing cost	14.83	16.97	16.97	16.97	16.97	16.97
Quantity	10,000	9,200	8,500	7,300	5,000	4,000
Profit	51,700	73,876	110,755	**131,619**	115,150	112,120

The index is the ratio of the value of each selling price of maximum profit divided by the values at the selling price of 20. These values are shown in Table 7.7.

The decision of which selling price and the routing alternative to use will result in maximum profit per period is computed with the aid of Table 7.7, as follows:

The first row indicates the candidate selling prices (the ones used in Table 7.6).
The second row is a summary of Table 7.5. It indicates which alternate process results in the maximum Relative Total Period Profit (RTPP) at each selling price.
The third row copies the maximum Relative Total Period Profit (RTPP) that refers to the alternate routing as indicated in the second line from Table 7.5.
The fourth row converts the Relative Total Period Profit to index numbers, assuming that the Relative Total Period Profit of the selling price of 20 is 100%. It is computed by dividing each relative value by 0.552—the RTPP of the selling price of 20.
The fifth row copies the relative quantity from Table 7.6.
The sixth row is the index result and is computed by multiplying the relative index (row 4) by relative quantity (row 5).

The decision is to choose the selling price and alternate process plan from the column that has the largest result index.

In the case of part "CROSS", as shown in Table 7.7, it is recommended to set the selling price to 35 and use the routing specified by alternative 9.

5.4 Testing the Algorithm

To test the validity of the proposed algorithm, the conjectured profit to be made by selling at each of the above selling prices is computed, as shown in Table 7.8.

For each selling price the selling quantity per period is given, the process alternative and the processing cost.

The profit is computed by computing the net profit per item.
Net Profit per item = Selling price—processing cost.
Total profit is computed by multiplying the quantity by net profit per item.

The selling price that results in the maximum profit is the recommended selling price, and the alternate process that caused it.

5.5 Summary of Maximum Profit

In this section we proposed a method of selecting an appropriate routing of a roadmap, and arriving at the optimum selling price. We also demonstrated that neither the minimum processing cost, nor the maximum production criteria of optimization, nor a higher selling price, is certain to result in maximum profit.

6 Determining Delivery Date and Cost

Traditional practice is that production management activities start with customer orders as input. Order delivery date is part of the order information. Delivery date plays a major role in the Production Management stages. Due date is one of the determining factors in establishing the quality of the shop's performance. However, the traditional system does not question how the delivery date was determined. In many cases, sales promises non-realistic delivery dates, but production management must regard them as a constraint.

The roadmap system may be used by sales and/or management to establish realistic delivery dates. Because routing is regarded as a variable, the processing lead time is flexible. The flexibility affects the processing cost as well.

Hence there is a direct relationship between the processing lead time (the delivery date) and the processing cost.

Solid information regarding this relationship might be very helpful to management while negotiating the selling conditions of price and delivery date with a customer.

The proposed module to establish realistic delivery dates is as follows: the last resource capacity plan will be referred to as present "load profile" (see Table 4.10). The load profile is a result of a finite capacity resource loading process. The loading method was described in Chap. 3. The load profile shows the existing jobs planned on all available resources.

In any plan there might be some resource idle time. When a new order arrives, its bill-of-material is used to compute the gross requirement of all items involved. Then the system searches inventory, computes the net requirement of each item and builds the working product structure. The matrix is called to generate a process plan to construct a time-based product structure. The time-based product structure is used to set the priorities of resource loading.

6 Determining Delivery Date and Cost

Loading the items of the new order, as computed by a working time-based product structure, will be determined by a method similar to the resource loading method described in Chap. 3. However, as it is a new order, it is superimposed on the present load profile. This means that the present load plan will remain unaltered and only the idle periods of each resource, or those occurring at the end of the loading periods, will be regarded as periods in which to load the new order.

However, as the new order is loaded only at these random times, the due date might be far ahead. The roadmap method may generate alternative processes, by using different resources. Since an alternate process cannot be as efficient as the best process, the cost of producing the order will increase. But on the other hand, the idle time of the alternate resources might be at an earlier period and thus the due date may become earlier.

The matrix method has several features that might be handy in this case.

One is the "Resource blocking option", where the matrix is asked to ignore certain resources.

The other is "Forced process planning option" where the resources to be used are dictated externally. These features are used to generate alternatives.

The alternative cost—delivery date will be generated in three steps:

1. Loading the new order and all its items as one unit with alternate process plans.
2. Generating alternatives while allowing working overtime or in shifts.
3. Splitting the order into several orders.

These methods will be demonstrated in this section.

The roadmap system can also be used when the customer insists on having the order ready by a certain delivery date but may compromise on the quantity. Sales personnel do not need to consult a process planner as they may generate alternatives by an appropriate computer program.

6.1 Generating Alternatives of Cost: Delivery Date: New Order

In this stage many alternatives will be generated and a table of cost-delivery date will be built. The alternatives will be generated following the working product structure and the priority set by the earliest start date of the low level items in a similar way as in the capacity loading as described in Chap. 3. The resource loading is superimposed on the last working load profile constructed.

The method is best demonstrated by continuing the example used in Chap. 4, Sect. 2.3.3.1.

Example: The company received a new order for 100 units of product #1 (see Table 4.10) and its matrix in Tables 4.3 and 4.4.

The customer would like to get an acceptable delivery date and cost for the order. Assume that not a single item is available in stock, and all the required items must

Table 7.9 Load profile for maximum production added order

Period	R1	R2	R3	R4	R5	R6	R7	R8	R9	R10	R11
1							90x	1,008	903	702	401
2							90x	903	907	702	401
3							90x	907	803	702	401
4							90x	1,004		702	401
5	301				602		80x	1,004		706	405
6	301		305		602	1,206	80x	807		70x	40x
7	301		305		602	506	80x			70x	40x
8	301		602			506	100x			70x	40x
9	301		205	602			100x			70x	40x
10	301		205		502		100x			70x	40x
11		201			502					70x	40x
12	30x	201			502					70x	
13	30x	201	502			60x					
14	30x	201	502			60x					
15	30x	201	502			60x					
16	30x	201	502			60x					
17	30x	201	502			60x					
18	30x					60x					
19	30x					60x					
20	20x	50x									
21	20x	50x									
22	20x	50x									
23	20x	50x									
24	20x	50x									
25	20x	50x									
26	20x	50x									
27	20x	50x									
28		50x									
29		50x									

be produced. The roadmap is called to generate a load profile including the new order. The last load profile is as shown in Table 4.10 while the new load order in Table 7.9. Items are distinguished by "x" as the last item code.

The delivery date can be determined by the load profile and the cost by computing the additional processing time on each resource and multiplying it by the resource unit cost.

6.1.1 Generating Alternatives for Cost-Delivery Date

Several methods are available to reduce lead time, such as:
- Working two or three shifts
- Work overtime
- Split order

6 Determining Delivery Date and Cost

Table 7.10 Load profile for splitting the order

Period	R1	R2	R3	R4	R5	R6	R7	R8	R9	R10	R11
1								1,008	903	702	401
2								903	907	702	401
3								907	803	702	401
4								1,004	90x	702	401
5	301				602			1,004	90x	706	405
6	301		305		602	1,206		807		7P1	4P1
7	301		305		602	506		90x		7P1	4P1
8	301		602			506		90x	80x	7P1	4P1
9	301		205	602		3P1			80x	7P1	4P2
10	301		205		502	3P1				7P2	4P2
11		201			502	3P1		80x		7P2	4P2
12	3P2	201			502	3P1		80x		7P2	
13	3P2	201	502		2P1	6P1		100x			
14	3P2	201	502		2P1	6P1		100x			
15	3P2	201	502		2P1	6P1		100x			
16	2P2	201	502			6P1					
17	2P2	201	502		5P1	6P2					
18	2P2		2P1		5P1	6P2					
19	2P2		2P1		5P1	6P2					
20		5P1	2P1			6P2					
21		5P1			5P2						
22		5P1			5P2						
23					5P2						
24			5P2								
25			5P2								
26			5P2								
27											

These options and others were generated by the roadmap system. The load profile for the case of splitting the orders into two orders of 50 items each is shown in Table 7.10. In this table the symbols that designate the new order split designate the second digit by the letter "P" and the last digit as the first or second batch. The first digit is as usual the item number.

6.2 Summary

Delivery dates play a major role in timing of all stages in production management. However, traditionally these dates are set by management without necessarily having a practical basis. It depends on sales promises and customer demands. In many cases promised delivery dates are impractical, and might cause severe losses to the company, such as loss of reputation and extra cost of meeting the promised dates by working overtime, extra shifts or on weekends and holidays.

Table 7.11 Summary of alternate cost-delivery dates

Alt.	Cost	Delivery period	Remarks
1	87.6	50	Minimum cost process plan
2	183.0	48	Maximum production process plan
3	166.0	38	Improved alternate 2
4	160.3	33	Improved alternate 3
5	134.0	33	Cost improvement of alternate 4
6	90.8	43	Improved alternate 1
7	91.4	39	Improved alternate 6
8	91.25	39	Two shifts for alternate 1
9	120.8	30	Splitting the order

Roadmap methods do not pretend to remedy the flaws in methods of establishing delivery dates; they merely aim at supplying management with accurate data on the cost—delivery date relationship. Management might be able to use such data in negotiations with the customer.

This section presented a method for computing the relationship between cost and delivery date for an order. It considers the available stock and the present load of the company. Several methods have been simulated; the results are presented in Table 7.11.

6.3 Chapter Summary

Managing of an enterprise is a decision making process. The technical data for decisions is generated and supplied by engineers. However, an engineer's criteria of optimization in making decisions, are not usually the same as those of management. The most common criterion of optimization is minimum cost or maximum production. Engineering decisions i.e. design and process planning, vary according to the criterion employed. Engineers are generally not economists or production management experts. They are experts in product design and process planning. However, they are required to make decisions on topics that are out of their line of expertise, and those decisions are transferred to management who, based on that data, implement sophisticated mathematical management decision models. Therefore, management decisions are restricted by their dependence on engineering data.

Implementing the concepts discussed in this book means that engineers should present alternatives and leave it to management to select, by its own criteria, the best solutions.

This section demonstrated such methods in three cases; one in resource planning, in determining sales prices that will maximize profit and in negotiation of delivery dates and cost.

Other applications such as company level of competitiveness, scheduling, and production management may benefit from the roadmap method. These three cases, and the others, are based on the roadmap method.

The roadmap provides means for generating routing in less than a second of elapsed time; therefore it can introduce new degrees of freedom and can treat the whole manufacturing process as one all-embracing dynamic system.

Index

A
Artificial, constraints, 8, 19, 36

B
Basic process (BP), 19, 24, 27, 74–76, 152–155, 159–161, 166
Bill-of-materials, 1, 4, 6, 8, 45, 57, 61, 66, 70, 74, 75, 80–85, 95, 119, 172
BP. *See* Basic process (BP)
Budget, 5, 10, 12, 132

C
Capacity planning, 5, 10, 13, 14, 36, 53, 59–63, 73, 74, 79, 80, 85–94, 96, 153, 167
Capital tied down in production, 2, 4, 6, 8, 13, 45, 62, 151
Cash flow profit forecasting, 12
Chain of activities, 4, 7, 8
Combinatorial problem, 33–36
Complex, 1, 3, 13, 15, 45, 63, 139, 143, 144, 146, 147
Computational data, 21, 22
Conceptual design, 9
Constraints, 1, 3, 4, 6, 8, 11, 18–20, 22–25, 27, 36, 43, 59, 102, 138, 144–148, 150, 153, 159, 172
Criteria of optimization, 2, 4, 19, 46–50, 52, 61, 62, 71, 89–93, 151, 152, 158, 166, 172, 176
Critical order, 13, 14, 53–55, 80–83, 85, 86

D
Decisions, 2, 18, 45, 73, 102, 133, 151–177
Decision support system (DSS), 11, 12, 151–177

Delivery dates, 1–5, 13, 14, 45, 46, 52–55, 57, 59, 60, 62, 63, 73, 80, 164, 167, 172–177
Design, 2, 18, 46, 74, 101–150, 157
Disruptions, 2, 6, 7, 13, 45, 46, 51, 53, 70, 73, 134, 156
DSS. *See* Decision support system (DSS)
Dynamic, 1, 2, 9, 19, 37, 38, 49–51, 53, 63, 70, 73, 99, 102, 177
Dynamic programming, 36–38
  Bellman theory, 36

E
Economic lot size, 12, 59, 168
Engineering drawings, 17, 18
  design, 11, 128
Engineering specifications, 9, 103

F
Finance, 4–6, 12, 151, 160
Flexibility, 2, 6, 14, 15, 46, 48–51, 53, 63, 64, 70, 71, 103, 111, 141, 145, 172
Free operation, 14, 50, 64, 65, 67–70, 95–98
Free resource, 14, 64, 65, 67, 70, 95, 98, 99

H
Human oriented activities, 2, 18, 152

I
Imaginary resource, 19, 153, 154, 158, 159, 166
Industrial management, 4–6, 12
Inventory, 3, 6, 10, 48, 55, 56, 63, 66, 112, 124, 152, 172

J

Job release, 13, 46, 52, 53, 62–63, 70, 74, 80, 94–95

L

Lead time, 2, 6, 8, 9, 11, 13, 18, 45, 51, 62, 89–94, 101, 112, 114, 125, 133, 164, 172, 174
Look ahead, 12, 14, 69, 96

M

Management, 2, 4–6, 9, 12, 13, 18, 45, 51–64, 66, 67, 73, 85, 101, 107, 108, 113, 151–177
Manufacturing,
Manufacturing process, 2–4, 7, 9, 45, 46, 63, 101, 113, 114, 152, 177
Master production plan, 12, 156, 157
Mathematical problem, 8, 19, 25, 36
Mathematical techniques, 8, 105
Matrix solution, 19, 36–44, 60, 61, 160
Maximum production, 2, 30, 34–36, 46–50, 52, 62, 71, 89–92, 162, 166–168, 172, 174, 176
Maximum profit, 12, 52, 61, 159, 166–172
Methods, 1–3, 5, 8–11, 13, 20, 27–30, 35, 36, 38, 43, 44, 46, 49–53, 55–61, 63, 65, 66, 70, 73, 89, 101, 105, 108–120, 123–127, 131, 132, 136, 138, 140, 142, 147, 148, 150, 152, 153, 156–159, 162, 166, 167, 172–174, 176, 177
Minimum cost, 30, 33–36, 48–50, 52, 71, 86–91, 93, 94, 105, 159, 166–168, 176

O

Operation
  optimization, 30, 34, 36
  transformation, 26–31
Optimization, 2, 4, 9, 11, 18, 19, 30, 33–36, 46–48, 52, 79, 85, 89, 94, 159, 166–168
Optimum, 4, 7–9, 11, 12, 17–19, 25, 26, 30, 34, 36–38, 40, 44, 51, 58, 102, 132, 153–155, 157, 169, 170, 172
Order delivery date, 14, 52, 54, 80, 172
Order logbook, 12

P

Part optimization, 30, 34, 36
Penalty, 30, 34–36, 65, 68, 80, 86, 88, 89, 96, 148, 153
Plant layout, 12

Priority code, 40, 42, 43, 160
Probability, 4, 5
Process
  planner, 4, 8, 11, 15, 18–21, 23, 30, 35, 52, 103, 152, 159, 166, 173
  planning, 5, 8–11, 14, 15, 17–44, 46, 47, 60, 64, 73, 85, 89–94, 101, 125, 153, 158, 159, 166–168, 173, 176
  planning task, 19
Product designer, 4, 8, 113, 138, 146, 148, 149
Production
  planning, 1–3, 6, 8, 12, 13, 15, 18, 49, 50, 52–53, 63, 64, 66, 73–99, 153, 155–156, 158
  planning and control, 1
Productivity, 2, 8, 9, 46, 48, 141, 142
Product
  selling price, 12, 168, 170
  structure, 3, 9, 11, 13, 14, 50, 52–62, 65, 85, 91, 124, 145, 154, 172, 173
Profitable, 15
Profits, 4, 6, 12, 52, 61, 151, 159, 164, 166–172, 176

R

Relationship code, 43
Resource
  planning, 5, 12, 18, 154, 156–166, 176
  selection, 25–26, 38, 41–43
Roadmap, 7, 11, 13, 15, 19, 51–63, 73, 75, 79, 125, 131, 152–155, 157–162, 166–168, 172–177
  concept, matrix, 19
  system, 13, 73, 131, 152, 157, 168, 172, 173, 175
Routing, 1, 2, 4, 6–8, 11, 13, 15, 17–44, 46–53, 60, 61, 63, 64, 73, 152, 157, 159, 166, 167, 170–172, 177
  process planning, 11, 15, 17–44, 46, 47, 60, 73, 166, 167

S

Sales, 6, 113, 145, 152, 153, 155, 156, 164, 167–173, 175, 176
Scheduling, 1, 3, 7, 13, 18, 35, 46, 48–50, 52, 63–67, 71, 94–96, 99, 152, 153, 167, 177
Sequence of decisions, 18, 19, 36
Shop floor, 6, 8, 12, 14, 46, 52, 53, 62, 66, 68, 70, 73, 81, 85, 86, 96, 103, 130, 153, 156
Shop floor control, 13, 14, 53, 63–71, 73, 74, 80, 95–99

Index 181

Simplicity, 6, 26, 59, 71, 74
Standardization and simplification, 6
Stock allocation, 13–14, 53–58, 74, 79–85

T
Technical knowledge, 25

Technological constraints, 19, 20, 23, 24
Technology stage, 19, 152

V
Value engineering, 6, 7